Advances in Geopolymer-Zeolite Composites - Synthesis and Characterization

Edited by Petrică Vizureanu and Pavel Krivenko

Published in London, United Kingdom

IntechOpen

Supporting open minds since 2005

Advances in Geopolymer-Zeolite Composites - Synthesis and Characterization
http://dx.doi.org/10.5772/intechopen.93360
Edited by Petrică Vizureanu and Pavel Krivenko

Contributors
Ahmed Nmiri, Tatiana Samarina, Esther Takaluoma, Outi Laatikainen, Muktikanta Panigrahi, Saralasrita Mohanty, Ratan Indu Ganguly, Radha Raman Dash, Petrică Vizureanu, Dumitru Doru Burduhos Nergis, Andrei Victor Sandu, Diana Petronela Burduhos Nergis, Madalina Simona Baltatu, Eduard Fron, Indhumathi Anbarasan, S. Nagan Soundarapandiyan, S.R. Sanjaiyan, Wega Trisunaryanti, Karna Wijaya, Desi Suryan, Uswatul Chasanah, Pavel Krivenko, Volodymyr Kyrychok

First published in London, United Kingdom, 2021 by IntechOpen
IntechOpen is the global imprint of INTECHOPEN LIMITED, registered in England and Wales, registration number: 11086078, 5 Princes Gate Court, London, SW7 2QJ, United Kingdom
Printed in Croatia

British Library Cataloguing-in-Publication Data
A catalogue record for this book is available from the British Library

Additional hard and PDF copies can be obtained from orders@intechopen.com

Advances in Geopolymer-Zeolite Composites - Synthesis and Characterization
Edited by Petrică Vizureanu and Pavel Krivenko
p. cm.
Print ISBN 978-1-83968-988-8
Online ISBN 978-1-83968-989-5
eBook (PDF) ISBN 978-1-83968-990-1

For EU product safety concerns:
IN TECH d.o.o., Prolaz Marije Krucifikse Kozulić 3, 51000 Rijeka, Croatia, info@intechopen.com or visit our website at intechopen.com.

Meet the editors

Petrică Vizureanu obtained an MSc and Ph.D. in Heating Equipment from Gheorghe Asachi Technical University, Romania, in 1992 and 1999, respectively. He is currently a Full Professor and Scientific Supervisor in Materials Engineering at the same university. His research interests include expert systems for heating system programming, computer-assisted design for heating equipment, heating equipment for materials processing, heat transfer, biomaterials, and geopolymers. Dr. Vizureanu has more than 150 papers Publications are visible in over 150 papers in international journals and conferences (proceedings) and 35 books to his credit. Hirsch index: 18.

Pavel Krivenko is currently a professor in the Department of Technology of Building Structures and Products and director of the Scientific Research Institute for Binders and Materials, Kyiv National University of Construction and Architecture, Ukraine. He graduated from the Dnepropetrovsk Civil Engineering Institute, Ukraine, with a diploma in Engineer in Building Materials Technology in 1961. He received his Ph.D. (Eng) from the Kyiv Civil Engineering Institute, Ukraine, in1971, and a DSc (Eng) from Kyiv Polytechnical Institute, Ukraine, in 1986. His research interests include the synthesis of alkali-activated cements and materials based on them. He has published more than 20 books, more than 1,000 papers in international journals and conference proceedings, and more than 500 patents. He is a principal investigator in several national and international projects in Ukraine and abroad. He is a member of three International Union of Laboratories and Experts in Construction Materials, Systems and Structures (RILEM) technical committees on alkali-activated materials, board member for three international journals, winner of the State Prize of Ukraine in the field of Science and Engineering (2001), and supervisor of five DSc (Eng) and twenty-three Ph.D. theses.

Contents

Preface

Strong technological development and increasing population over the past few decades have led not only to a dramatic increase in the demand for raw materials and energy but also to the need for recycling and process intensification. Since recycling is seen as the most promising and cost-effective option to reduce the use of fossil energy and non-regenerable raw materials, in industry and oxidic materials applications, geopolymers seem to meet all the criteria to replace conventional materials and contribute to a sustainable future. By applying innovative technologies, the coefficient of use of waste for developing new materials can be increased simultaneously along with a decrease in carbon footprint and ecological impact. Accordingly, these eco-friendly materials will participate in cutting-edge research and applications due to their tailored properties, which include superabsorbent capacity, heavy metals encapsulation, flame retardancy, mechanical performance, electrokinetic behaviour, corrosion resistance, and thermal properties.

Geopolymer–zeolite composites and zeolite-like geopolymers are two different categories of adsorptive materials that have recently attracted increased interest. Geopolymer–zeolite composites are hybrid materials and unite the advantages of both constituents. The geopolymer serves as durable support, while the zeolite provides a high surface area, porosity, and adsorption capacity. The report clearly shows the beneficial influence of the use of zeolitic tuff in the starting mixture on the microstructure and adsorption potential of geopolymers.

This book joins activities and knowledge of researchers from multiple fields to present a comprehensive overview of the advances in synthesis and characterization of geopolymers, including base chemistry concepts, nanoscale characterization, and applications in top-level industry. It is organized into two sections on "Geopolymers" and "Zeolites" and includes eight chapters containing information about the theoretical approach to geopolymers and their applications in civil engineering, medicine, and other areas.

Chapter 1 discusses the regeneration of exhausted materials and available resource recovery options that the regeneration approach opens. It describes new forms of geopolymer adsorbents such as foams or core-shell structures and provides a short economic evaluation of resource recovery models.

Nowadays, globalization generates large amounts of waste that significantly affects storage areas and the surrounding environment. At the same time, the civil engineering sector is experiencing an exponential development process, which increases the demand for building materials and usable space. Therefore, the need to obtain new materials with lower exploitation costs and natural resources consumption became primary. Chapter 2 describes one solution that has been intensively studied in the past year, especially in this sector, consisting of the development of environmentally friendly materials through a mechanism called geopolymerisation.

Geopolymers are being studied extensively due to their potential applications in the construction sector. Chapter 3 is a study of a variety of geopolymer ingredients, including fly ash, bottom ash, high-carbon ferrochrome (HCFC) slag, sodium hydroxide, sodium silicate, plasticizer, and others that are used to synthesize geopolymers with improved mechanical properties such as compressive strength. Geopolymers are developed from HCFC slag by treating with sodium silicate, sodium hydroxide, and water-soluble plasticizers.

Geopolymer concrete (GPC) has significant potential as a more sustainable alternative for ordinary Portland cement concrete. Chapter 4 explains how GPC was introduced to reduce carbon footprints and thereby safeguard the environment. This emerging eco-friendly construction product finds application in precast and prefabricated structures due to the special curing conditions required. Sustained research efforts are being undertaken to make the product suitable for in situ applications. The developed technology will certainly address issues of huge energy consumption as well as reduce water usage.

Several new technologies are emerging to help achieve the aim of reducing energy usage in building sectors, eliminating greenhouse gas emissions, and recycling waste. Chapter 5 describes some of these technologies, including the development of a geopolymer binder that may be used as an alternative to ordinary cement Portland, the adoption of three-dimensional (3D) printing methods in civil engineering, and the integration of phase-change materials (PCMs) in cementitious materials to increase the energy efficiency of buildings. Most investigations in this area focus on the addition of microencapsulated PCM (MPCM) to standard concrete recipes.

Chapter 6 discusses zeolite-like analogies to natural aluminosilicate minerals that may be synthesized in the cement matrices of the Na(K)-Al_2O_3-SiO_2-H_2O system. The structure formation and properties of the alkali-activated aluminosilicate cement-based materials at low temperatures may be regulated by changing cement composition, curing conditions, type and concentration of the alkaline activator solution, and solution-to-solid ratio. Directed regulation of the low-temperature structure formation process is a key instrument allowing to obtain a wide range of special materials using various types of alkali-activated aluminosilicate cement and curing conditions.

Chapter 7 critically reviews the studies related to structural and photophysical properties of metal clusters within zeolite matrices and summarizes the progress made in understanding host-guest interactions. The goal is to provide useful insight into the nature of such interactions and experiments used in identifying the excited-state dynamics and reaction mechanisms leading to the emitting species. Especially interesting are the combined experimental and computational approaches used to elucidate the structures and electronic transition of clusters inside the cavity.

The last chapter explains the three-dimensional structure of zeolites, composed of AlO_4 and SiO_4, which are related to each other by sharing electrons from oxygen and are arranged tetrahedrally. Zeolites have been widely used in industrial processes as environmentally friendly heterogeneous catalysts, for ion exchange, and as adsorbents due to their high specific surface area, large pore volume, uniform micropore channels, and excellent thermal and hydrothermal stability. The use of zeolite as a catalyst in various industries is limited due to its narrow pores.

Petrica Vizureanu
The "Gheorghe Asachi" Technical University,
Iasi, Romania

Pavel Krivenko
Kyiv National University of Construction and Architecture,
Kyiv, Ukraine

Section 1

Geopolymers

Chapter 1

Geopolymers and Alkali-Activated Materials for Wastewater Treatment Applications and Valorization of Industrial Side Streams

Tatiana Samarina, Esther Takaluoma and Outi Laatikainen

Abstract

The EU has the ambitious goal to transition from linear to circular economy. In circular economy, the old saying of "one's waste is the other's treasure" is being implemented. In this chapter, valorisation of industrial side streams, traditionally branded as waste, is discussed with respect to their applications as raw materials for new adsorptive products – geopolymers (GP) and alkali-activated materials (AAM) – as adsorbents in wastewater treatment. The chemical nature and structure of materials generally have great influence on GP/AAM adsorption capability. The approaches used for the raw materials preparation (chemical or physical) prior geopolymerization to increase the adsorption capacity of the final products will be discussed. Adsorption properties and performance of GPs/AAMs towards various contaminants are described, and the latest research on testing those materials as water remediation are reviewed. Special attention is paid to regeneration of exhausted materials and available resource recovery options that the regeneration approach opens. New forms of geopolymer adsorbent such as foams or core-shell structures are described and in the last part of the chapter, a short economic evaluation of resource recovery models is provided.

Keywords: adsorption, metal removal, nutrient recovery, geopolymer composite, wastewater treatment

1. Introduction

The United Nations have ratified 17 goals of sustainable development, of which responsible consumption and production is directly, while economy, innovative industry, infrastructure, and climate action are indirectly related to circular economy and the need of sustainable production [1]. Sustainability in the processing industries can be applied along the main value chain, e.g. from metal extraction to metal recycling, but can also be applied to the associated waste materials. Copper and iron mining alone are estimated to generate yearly about 5 bn tons of tailings [2], i.e. the fraction of the processed ore, after extraction of the valuable minerals. Finding a way to successfully reuse vast amount of this material and other waste sources is a great step towards circular economy.

While recycling an initial waste or side-stream, the material can be upcycled, meaning the newly derived product is of higher intrinsic value and properties or downcycled, where the new material has lower value. A prominent example for downcycling is the reuse of plastic bottles as fleece and carpet material. The material has less intrinsic value, because the carbon chains of the plastic polymer are shortened. Geopolymerization of industrial side streams is an upcycling process, as the geopolymer (GP), utilized as concrete and binder, has a higher value than the initial industrial by-products. To obtain upcycling, energy is put into the system, however, since tailings have usually been milled, the material has already undergone energy intensive steps and can therefore readily be used as starting blocks for geopolymerization. Utilizing tailings for upcycling into GPs, is therefore beneficially in terms of waste management, process energy, and emission of greenhouse gases, as the energy used in the beneficiation process is passed onwards into the geopolymerization process. Upcycling often requires further energy sources to achieve higher valuable material. In geopolymeration or alkaline activation this means the addition of chemicals, and in some occasions, such as analcime tailings, the addition of thermal energy [3].

In this chapter the valorization of high volume, inorganic side streams from mining, chemical industries, steel processing, and waste incineration into new adsorbents useable for water treatment is discussed. The purpose is to show how the material undergoes value change from side stream to potentially highly functional material.

As every tailing and every ash has a different chemical and mineral composition, tailoring of the properties of resulting adsorbents is possible by careful choosing of precursor materials. Aluminosilicates form the backbone of the geopolymer structure, but ion exchange, channel size, and physical properties are affected by the minerals used for geopolymerization [4–6]. Lastly, by controlling of the geopolymerization conditions, also the macroscopic structures can be developed by using various manufacturing methods from foaming to granulation.

The ultimate goal of using GPs/AAMs in water purification is to be able to recover valuable materials such as nutrients or battery chemical metals from contaminant-rich wastewater streams. In other words, the target is to use one industrial side stream to recovery of valuable material from another side stream or waste water in order to multiply circular economy potential.

2. Raw materials and preparation of GP/AAM materials for water treatment applications

This section summarizes different types of aluminosilicate precursors, occuring naturally or derived from industrial processes. Materials, which are currently abundant and/or urgent to dispose of, fall within the ambit of the section, but cover only water treatment applications not geopolymer production for construction industry, e.g. substitutes for Portland cement or as tailings' covering.

2.1 Conventional and new sources of aluminosilicate precursors for GP/AAM preparation

Ashes. *Fly ashes (FA)* are prominent materials used as alkaline-activation binders. FAs are abundant yet complex materials, the composition of which is affected by co-incinerated material. Produced mostly by coal-fired electric and steam generating plants, coal FAs represent the greater part of generated FAs with the estimated flow of approximately 750 Mt. in 2015. The utilization rates of FAs differ greatly worldwide. Thus, rates in the developed countries equal 96%, 85%, 50% and 65% from produced streams for Japan, Germany, UK and USA, respectively [7]. For the developing countries with growing economics such India and China, the utilization rates are 38% and 45%, yet rate 66% for Asia in general are reported. Russian Federation and Africa implement FA as secondary resource less than 20% [7].

The composition of FA varies widely as it is derived initially from various primary sources: municipal waste/sludge co-incineration, different coal types, or subspecialized byproducts from industrial treatment plant (paper, forestry industry or agriculture). The combustion and cooling processes have profound impact on the characteristics of FA (particles size, shape, surface area, uniformity, etc.) as well as its composition and impurities' inclusion.

Mainly, ASTM C 618 specification is applied to indicate the class of FA used for geopolymer preparation; however, a local/field or an unspecified labelling is also common. Coal FA (class F [8, 9] and C [10]) has been extensively considered as an aluminosilicate source for GP production, while the exploitation of biomass and co-incinerated FAs is less common [11, 12]. On the other hand, the utilization of these FAs particularly in the GP production for water treatment sector might be also beneficial. It would reduce the FA accumulation in landfills, and improve adsorbents' LCA in comparison with metakaolin-based GPs.

Although FAs were studied as adsorptive materials previously [7, 13, 14], concerns on potential toxicity of impurities and convenience of use have encouraged to seek more suitable forms of FA-based materials for water treatment sector.

Municipal waste incineration bottom ash (IBA) has been traditionally considered as solid waste [15, 16]. IBA is mainly composed of Si, Al, Ca, and Na oxides, and could be classified as a hazardous or non-hazardous waste depending on the amount of toxic metal(oid)s. IBA's main applications are in the engineering field as secondary materials in form of weathered bottom ash (after outdoor ageing for 2–3 months for pH stabilization). However, new applications of IBA have emerged in recent years as use as an aluminosilicate source for GPs/AAMs [17] including adsorbents [15, 18–20]. IBA by itself also was investigated as an adsorptive material for metal removal [21, 22]. In most of the studies, mixtures of IBA with various aluminosilicate precursors (BFS, FA, metakaolin) are used in order to obtain desired mechanical characteristics [23]. For the water treatment applications, IBA as raw material for GPs could have hidden benefits as the aluminum presented in it reacts with the alkaline activator and forms hydrogen gas, which leads to an increase in porosity [24]. Moreover, since a compressive strength of the resulting materials could be lower than for construction applications, the high porosity of materials and *in situ* stabilization of concomitant hazardous impurities via encapsulation could be attractive options [9, 19].

Pre-treatment of FAs and IBAs with various chemicals were suggested in order to reduce their toxicity and to meet the environment requirements of pristine materials or/and GPs/AAMs based on them [14, 25–27].

Steel industry waste. *Blast furnace slag (BFS), dust, and sludge.* BFS is another copious industrial nonmetallic by-product that is used widely as GP precursor in civil engineering [28]. Similar to ashes, steel industry wastes are mainly composed of Si and Al oxides, while Mg and Ca oxides could consist up to 35–60% of the material by weight [29]. Iron and sulfur are the major impurities in BFS, derived from the iron-smelting process. Ground granulated BFS has a high specific area due to the small particle size distribution, which makes it an excellent candidate as adsorptive material [13, 30]. However, in order to avoid the leaching of heavy metals from BFS during deployment of these materials in water remediation techniques, alkaline activation were suggested for entrapment/binding impurities within the GPs' matrix [31–34]. The presence of significant amounts of silica, aluminosilicates, and calcium-alumina-silicates in a pristine material makes the geopolymerization process rapid and effective resulting in rigid and enduring compositions. Thus, BFS was used to enhance the stability of metakaolin- and FA-based GPs [35].

Basic oxygen and electric arc furnace slag. These materials are sub-categorized, depending on the process of their formation. Both basic oxygen furnace slag (BOFS) and electric arc furnace slag (EAFS) are formed during the steelmaking process [36, 37]. These types of slags are similar in composition to BFS, except for their iron, manganese, chromium and sulfur contents, which are higher in BOFS and EAFS. EAFS was modified by alkali activation [38] and its adsorption properties towards copper were compared to raw EAFS. Significant amount of posnjakite were detected in the crystalline phase after adsorption of copper that could explain the drastically high removal efficiency of this AAM.

The accumulation of BOFS has become a significant issue due to its generation in large quantity, high disposal costs, and unsuitableness in cement industry due to high iron oxide content. Sarkar et al. adopted BOFS as a raw material for obtaining of GPs and investigated Ni^{2+} [39], Zn^{2+} [40], and F^{-} [41] removal. BOFS was used by Sithole et al. as a precursor for AAM preparation [42, 43]. In order to achieve highly porous structures for percolation column tests, a foaming agent (hydrogen peroxide) was added.

Red mud, silica fume and ore materials. *Bauxite* is a sedimentary rock, rich in aluminum oxide minerals and accompanied by kaolinite, quartz, and iron oxides. The amount of impurities varies depending on the place of origin. Bauxite itself has recently been tested in water treatment applications for purification of fluor- and arsenic-contaminated waters [44, 45]. However, even keener interest is observed in the valorization of bauxite residues, rich in iron and aluminum, as GPs/AAMs [46] and their water purification applications [47, 48].

Alumina manufactured through Bayer process from bauxite mainly goes to aluminum metal production. The rest (up to 10% of the whole production flow) is used as fillers in construction, in glass production, abrasive materials, and catalysts. Depending on the field of application, aluminum oxide could be called alumina, aloxide, aloxite, or alundum, and their respective waste materials would conform to these terms. Aloxite, for instance, is used as catalyst and/or catalyst support in organic chemistry due to its physicochemical stability and unique surface properties. As waste material, it could be accumulated in immense amounts from gas purification, decolorization, and catalytic processes as well as refining and desulfurization of petroleum oils and waxes. One of the ways of its valorization is in the design of eco-friendly GPs [49] and adsorbents for wastewater treatment [50, 51]. Thus, the addition of aloxite to analcime before geopolymerization showed an increase in the specific surface area and pore volume [50].

Silica fume is an amorphous form of silicon dioxide, collected as a by-product of the silicon and ferrosilicon alloy production. Historically, the main field of application is as pozzolanic material for high performance concrete of high strength and low porosity, though its applications for designing of foamed GPs and immobilization of cesium are reported [52, 53].

Mine tailings are mining and mineral wastes. The proper disposal of tailings has gotten under strong scrutiny for environmental preservation during the last decades. In many cases, the tailings are fine particles, containing silica together with iron oxides, alumina, and other minor minerals. This constitution makes tailings an excellent source of material for GPs. Iron ore tailings were mixed with FA to produce GP for copper removal [54], while gold mine tailings with Al_2O_3 additive were the source for GP production used for lead removal [51]. Prophyllite waste materials obtained from mine were converted to prophyllite GPs by Panda et al. [55], and tested as adsorbent for Co, Cd, Ni, and Pb removal from model solutions. Magnesite tailings from talcum mines show a reactivity dependent on the calcination temperature: light burnt (700 – 1000°C), hard burnt (1000 – 1500°C), and dead burnt (1500 – 2000°C). For use in water treatment and as a GP precursor, a light burnt magnesite in the form of periclase MgO is suitable.

Natural materials: zeolites, clays, sedimentary rocks. Kaolin is a rock rich in kaolinite, a clay mineral, and the source of production for the most widely used GP precursor called metakaolin. Metakaolin, a disordered, activated, and dehydroxylated form of kaolinite, is obtained through calcination of kaolinite over 600°C. The temperature of calcination has direct impact on reactivity of metakaolin, and as a consequence, on the crystallinity of the produced GPs [56]. However, raw kaolin/kaolinite were also used in GP production [57], yet for water treatment applications just recently [58]. Moreover, valorization of spent metakaolin could be beneficial and decline the cost of metakaolin-based GP [59]. Other clay materials have also been utilized recently for GP design [60]. Thus, bentonite clay calcined at 700–800°C were used by Maleki et al. [61] for obtaining a magnetic GP for heavy metal removal. Laterite clay-based GP were proposed by Ghani et al. [62] as a promising adsorptive material for Ni and Co removal. Laterite was activated at 900°C prior the geopolymerization. Volcanic tuff is another naturally available material with high porosity and with a high potential for ion-exchange. It was used in [63, 64] as an abundant yet low-cost material for GP production, and subsequent Zn removal from water.

Zeolites and zeolitic materials are well-known microporous materials. Found in nature or obtained through synthetic procedure, they are considered to be selective adsorbents [65–67], catalysts [68], carriers in biotreatment [69] due to their unique structure. Although, naturally occurred zeolites are readily available, they generally show lower surface area than synthetic ones.

Recently, much attention has also been paid on how zeolite could be synthesized from low-cost materials [70]. GP-zeolite composites and zeolite-like GPs are two different categories of adsorptive materials, which have recently attracted increased interest [71]. GP-zeolite composites are hybrid materials, unite the advantages of both constituents. The GP here serves as a durable support, while the zeolite provides a high surface area, porosity, and adsorption capacity. For instance, metakaolinite–zeolitic tuff GPs have been proposed in [72]. The report clearly showed the beneficial influence of the zeolitic tuff addition into a starting mixture on the microstructure and the adsorption potential of GPs. Andrejkovičová et al. [4] prepared metakaolin-based GPs blended with by 25, 50 and 75% of Nižný Hrabovec zeolite. It was shown that the zeolite particles are responsible for the higher amount of crystalline phases, producing a more compact and firm microstructure of blended GPs. The amount of blender has significant influence on the order of adsorbed metals and on the adsorption capacities of the formulations. Hayashi et al. [63] incorporated clinoptilolite into GPs though sol–gel protocol in order to further use of the resulting coatings for heavy metal ion adsorption.

It should be noted that zeolitic phase could be incorporated into GPs' structures not only externally. Zeolite-like crystalline phases could be derived from synthesis routes through fusion method or even at moderate temperatures leading to zeolite-like GP structures. Javadian et al. [64] converted FA into a mesoporous aluminosilicate adsorbent through a fusion method at 600°C. Deng et al. showed that a hydrothermal synthesis of zeolite-like materials from IBA with higher crystallinity than through a fusion method is possible [73]. Similarly, Visa [74] converted FA into zeolite through a hydrothermal process. Rios et al. synthesized zeolite-like GPs from metakaoline at 100°C through the hydrothermal procedure [75]. Studies reported indicate that such materials have higher surface area and porosity than GPs/AAMs obtained through simple alkaline activation. Although the ultimate set of preferable conditions to form a GP instead of a zeolite are still under discussion, ratios Si:Al > 1.5 have been empirically established as providing more amorphous structures [60].

α-Analcime is a reject from spodumene refining at a Finnish lithium hydroxide plant, currently in piloting stage, and estimated to start the production in 2024, but is also found as a natural zeolite [3]. The small cavity size of analcime facilitates ion-exchange only for small mono- and divalent cations such as ammonium and Cu^{2+}, and also K^+, Ag^+, Tl^+, Rb^+ (at elevated temperatures), and with low adsorption capacity. Raw analcime is inert to alkaline activation and analcime requires either chemical activation by 3–5 M H_2SO_4 or thermal activation above 700°C [3].

Not infrequently, industrial side streams cannot be used alone for geopolymerisation due to disharmonious Si/Al molar ratios. Therefore, by-products are commonly used as mixtures of aluminosilicate sources [76]. **Table 1** summarizes the studies on different compositions of GPs/AAM that have been proposed for water and wastewater treatment applications. An afford was made to collect and match the precursors, synthetic protocol specificity, and distinctive characteristics resulting materials.

2.2 Forms and manufacturing techniques of GPs/AAMs for water treatment

Originally, a basic composition applied for manufacturing GP/AAM adsorbents consisted of an alumosilicate precursor, an alkali, and an additional source of silicate in a form of water glass. Initially, both sodium and potassium forms of alkaline activators were used to induce geopolymerization. In the vast majority of the research reviewed, sodium alkaline and water glass are used in the activation process. It was shown by Bakharev that dissolution rates of the minerals was higher when a sodium form is used [148]. Luukkonen et al. [149] found that adsorption characteristics of metakaolin-based GP prepared with NaOH is better than with KOH in case of ammonium removal. An in-depth discussion of G chemistry and vivid explanations could be found in the latest reviews [57, 150, 151].

Forms and manufacturing techniques of GPs/AAMs for water treatment application are emerging and evolving constantly. In the first instance, *powdered forms* were mostly used for gaining of adsorption characteristics of materials. At first, GPs/AAMs were manufactured in bulk forms to be crashed to powder or rubbles after curing procedures. However, these materials have relatively low porosity, and addition of foaming agents were appealing for increasing surface area, pore volume, and porosity. In **Table 1** forms of GPs/AAMs reported and specific additives listed along aside with their synthetic procedures and properties gained. For bulk samples, species are usually sealed with plastic films to prevent moisture evaporation since the presence of water increases porosity. Curing and aging are usually carried out at temperatures ranging from 20 to 105°C. Commonly, an industrial application of powdered forms requires instance pressure filtration and an additional separation step after adsorbent exhaustion. Both these processes increase a cost of treatment, its complexity, limiting a regeneration ability causing sludge accumulation.

Granular forms of GPs/AAMs are more preferable for large-scale applications. FA-based GPs supported on inert substrate were proposed with the aim of overcome these limitations [107]. A simple technique similar to the conventional GP preparation was applied to design floatable light granules for phosphorous removal. Moreover, spherical granules could be produced by *in situ* geopolymerization during granulation by a high-sheer granulator, where a liquid alkaline activator acts as a binding liquid [59, 152]. A granule size distribution is a function of a liquid to solid ratio, granulation time, and a rotation speed. While the amount of liquid required

GP/AAM	Precursor/additives	Preparation method, prime oxide ratios	Surface Area/Pore Volume/Pore size	Type/form of GP	Ref.
MK-GP with TiO_2	MK	HT	27.21 m^2/g 0.207cm^3/g 2.19 nm	Bulk	[77]
MK-GP	MK	SSM	53.95 m^2/g 0.061 mL/g 5.38 nm	Porous/Spheres, 2–4 mm	[78]
MK-GP	MK	AA Si/Al = 1.7	–	Bulk	[79]
MK/Z-GP	MK Zeolitic tuff	AA SiO_2/Al_2O_3 = 1	–	Bulk/Discs	[72]
MK-GP	MK	AA Si/Al =	12.21 m^2/g 0.037cm^3/g	Bulk	[80]
MK-GP	MK SDS 0.06 wt%	SSM SiO_2/Al_2O_3 = 1.6	53.95 m^2/g 1.29 cm^3/g 15 nm	Porous/spheres 2–4 mm	[81]
MK/FA-GP	2/3 MK 1/3 bioFA (w/w)	AA SiO_2/Al_2O_3 H_2O_2	–	Porous/Monolith	[11]
MK-GP	MK	AA foaming, SiO_2/Al_2O_3 = 5	–	Foam/Powder <100 μm	[82]
MK-GP/alginate hybrid	MK sodium alginate	AA + SSM SiO_2/Al_2O_3 = 1.6	16.2 m^2/g 0.05 mL/g 11.5 nm	Bulk/Spheres 2–4 mm	[83]
MK-GP	MK	AA SiO_2/Al_2O_3 = 3.2	39.24 m^2/g	Bulk/Powder 150 μm	[84]
MK-GP functionalized with CTAB	MK silica fume	AA CTAB	216 m^2/g 0.22 cm^3/g	Bulk/rubbles, 1.5 mm	[85]
MK/FA-GP	MK:FA 2:1 wt	AA	7.9 m^2/g	Porous/discs	[86]

GP/AAM	Precursor/additives	Preparation method, prime oxide ratios	Surface Area/Pore Volume/Pore size	Type/form of GP	Ref.
MK-GP	waste MK	AA SiO_2/Al_2O_3 = 1.5	–	Bulk/Powder, granules	[59]
MK-GP	MK	AA SiO_2/Al_2O_3 = 3.2	39.24 m^2/g	Bulk/Powder, 150 μm	[87]
MK-GP activated with hull ash	MK BioFA TiO_2	SSM SiO_2/Al_2O_3 = 3.18*	–	Porous/Spheres 2–3 mm	[88]
MK/FA-GP	MK:FA 50:50 wt%	SSM SDS	–	Foam/ Spheres	[89]
MK-GP/alginate-chitosan hybrid	MK alginate/chitosan	SSM AA 0.5 wt% H_2O_2, 1.5 wt% SDS	230 m^2/g 0.99 mL/g 35 μm	Porous/Spheres	[90]
MK/FA-GP	MK FA class C 60:40 wt%	AA SiO_2/Al_2O_3 = 2.7	–	Bulk/Powder 63–125 μm	[91]
MK/BFS-GP	MK BFS 60:40 wt%	AA SiO_2/Al_2O_3 = 3.1	–	Bulk/Powder 63–125 μm	[91]
MK-GP	MK	AA SiO_2/Al_2O_3 = 2.31**	–	Bulk/Powder, 200 μm	[92]
MK-GP/ coal gangue hybrid	MK gangue 50/50 wt%	AA SiO_2/Al_2O_3 = 4.0	26.41 m^2/g 0.330 cm^3/g	Bulk	[93]
MK-GP	MK	AA SiO_2/Al_2O_3 = 2–8	–	Bulk/Pervious	[94]
MK-GP	Waste MK Aluminum scrap recycling waste 1:1 (w/w)	AA SiO_2/Al_2O_3 = 1.25**	15.95 m^2/g	Bulk/Granules 4–11.2 mm	[95]

GP/AAM	Precursor/additives	Preparation method, prime oxide ratios	Surface Area/Pore Volume/Pore size	Type/form of GP	Ref.
MK-GP	MK	AA	$8.16\ m^2/g$ $0.021\ cm^3/g$ 10.5 nm	Bulk/ Granules 0.5 mm	[96]
MK-GP functionalized with HDTMABr	MK	AA HDTMABr	–	Bulk/Powder, 53 μm	[97]
MK-GP MK/aloxid ANA-GP ANA/aloxid-GP	MK analcime aluminum oxide	AA** MK-GP $SiO_2/Al_2O_3 = 3.96$ MK/aloxid GP $SiO_2/Al_2O_3 = 2.13$ ANA-GP $SiO_2/Al_2O_3 = 7.01$ ANA/aloxid-GP $SiO_2/Al_2O_3 = 3.60$	MK-GP $19.97\ m^2/g$ $0.131\ cm^3/g$ 26.24 nm MK/aloxid GP $6.36\ m^2/g$ $0.036\ cm^3/g$ 23.18 nm ANA-GP $0.69\ m^2/g$ $0.003\ cm^3/g$ 21.69 nm ANA/aloxid-GP $38.29\ m^2/g$ $0.125\ cm^3/g$ 13.07 nm	Bulk/Powder, 63–125 μm	[50]
MK/Z-GP	25% MK 75% zeolite	AA $SiO_2/Al_2O_3 = 1$	$57.5\ m^2/g$	Bulk	[4]
MK-GP functionalized CTAB	Calcinated halloysite clay	Precipitation $SiO_2/Al_2O_3 = 2.91$ $CTAB/Cu_2O/TiO_2$	$34.8\ m^2/g$ 29.7 nm	Bulk	[98]
MK-GP functionalized with CTAB	MK	AA CTAB	$26.45\ m^2/g$ $0.121\ cm^3/g$ 9.12 nm	Bulk/Powder, 125 μm	[99]
MK-GP, magnetic hybride	MK Magnetite 5 wt%	AA $SiO_2/Al_2O_3 = 4.55^*\ H_2O_2$	$19.5\ m^2/g$ $0.045 cm^3/g$ 10.4 nm	Porous	[100]

GP/AAM	Precursor/additives	Preparation method, prime oxide ratios	Surface Area/Pore Volume/Pore size	Type/form of GP	Ref.
MK/Silica-GP functionalized with Cr	MK silica fume 9:1 (w/w)	AA SiO_2/Al_2O_3 = 1.90**	30 nm	Bulk/Membrane	[101]
MK-GP	MK	AA SiO_2/Al_2O_3 = 2.14**	–	Bulk/Powder, 355 μm	[102]
MK/FA-GP	MK FA	AA SiO_2/Al_2O_3 = 2.45**	27 m^2/g	Bulk/rubbl 1.0–0.3 mm	[103]
MK-GP	MK	AA SiO_2/Al_2O_3 = 4.0	21 m^2/g 1252 mm^3/g 0.32 μm	Bulk/Granules, 3 mm	[104]
MK-GP alginate hybrid	MK sodium alginate TiO_2	SSM $SiO_2:Al_2O_3$ = 4	20 m^2/g 714 mm^3/g 0.11 μm	Bulk/Spheres, 2 mm	[105]
MK-GP/ magnetic hybrid	MK	AA SiO_2/Al_2O_3 = 4.55 H_2O_2	42.92 m^2/g 0.052cm^3/g 4.88 nm	Porous	[106]
MK-GP/LECA	MK LECA support	AA SiO_2/Al_2O_3 = 1.5	–	Bulk/Granules, 4–8 mm	[107]
MK/Biochar-GP	MK Biochar	AA H_2O_2	37.46 m^2/g	Foam/membrane	[108]
MK-GP functionalized with $K_4Fe(CN)_6$	MK	AA SiO_2/Al_2O_3 = 3.60 H_2O_2	35 m^2/g 55 cm^3/g	Foam	[109]
MK-GP/graphene oxide hybrid	MK graphene oxide 10 wt%	AA SiO_2/Al_2O_3 = 0.45	–	Bulk/ Particles, < 0.5 mm	[110]
FA-GP	Fly ash	AA SiO_2/Al_2O_3 = 2.03	–	Bulk	[111]

GP/AAM	Precursor/additives	Preparation method, prime oxide ratios	Surface Area/Pore Volume/Pore size	Type/form of GP	Ref.
FA-GP	Fly ash, 75 μm	FM $SiO_2/Al_2O_3 = 1.98^*$	8.22 m^2/g 2.9 nm	Bulk/Powder	[64]
FA-GP Iron-enriched	Calcinated FA, < 70 μm	FM $SiO_2/Al_2O_3 = 1.00$ $Fe_2O_3/Al_2O_3 = 0.151$	–	Bulk	[112]
FA-GP modified with iron	Coal fly ash	AA $SiO2/Al2O_3 = 1.43^*$	162.38 m^2/g 0.126 cm3/g 3.90 nm	Bulk/Powder	[113]
FA-GP	Fly ash	AA $SiO_2/Al_2O_3 = 4.61$	–	Bulk/Powder, 71–90 μm	[9]
FA/IOT -GP	Fly ash IOT 70:30 (w/w)	AA H_2O_2	6 nm - 360 μm	Porous/Cubes	[54]
FA-GP	Fly ash	AA $SiO_2/Al_2O_3 = 1.12^{**}$	20.48 m^2/g 19.62 nm 0.070 cm^3/g	Bulk/Powder, 74 μm	[114]
FA/Z-GP	Fly ash Fajustite	HT $SiO_2/Al_2O_3 = 0.69^{**}$	174.35 m^2/g 0.14 cm^3/g 9.69 nm	Bulk/Powder, 74 μm	[114]
FA-GP	Fly ash C	AA $SiO_2/Al_2O_3 = 3$	–	Bulk/Powder	[115]
FA-GP/LECA	Fly ash C LECA support	AA $SiO_2/Al_2O_3 = 1.5$	–	Bulk/Granules, 4–8 mm	[107]
FA-GP	Fly ash	AA $SiO_2/Al_2O_3 = 5.36$	–	Bulk/Powder, 71–90 μm	[116]
FA-GP	Fly ash	HT	–	Bulk	[117]
FA/Z-GP	Fly ash BFS 4:1 (w/w)	HT $SiO_2/Al_2O_3 = 3.49^*$	–	Bulk/Powder	[118]

GP/AAM	Precursor/additives	Preparation method, prime oxide ratios	Surface Area/Pore Volume/Pore size	Type/form of GP	Ref.
Fly ash/ iron oxide hybrid	Fly ash Fe_2O_3 5 wt%	AA $SiO_2/Al_2O_3 = 3.30^{**}$	60.75 m^2/g	Bulk/Powder, 50 μm	[119]
FA-GP/Graphene hybrid	Fly ash graphene (1 wt%)	AA $SiO_2/Al_2O_3 = 3.41^{**}$	20.41 m2/g 0.047 mL/g 9.73 nm	Bulk	[120]
FA/BFS-GP	Fly ash BFS	HT $SiO_2/Al_2O_3 = 3.23^{*}$	76.6 m^2/g 0.24 cm^3/g 12.5 nm	Bulk	[35]
FA-GP	Boiler fly ash < 80 mesh	HT $SiO_2/Al_2O_3 = 2.75^{**}$	27.51 m^2/g 0.032 mL/g	Bulk	[121]
FA-GP	Fly ash	AA $SiO_2/Al_2O_3 = 3$	29 m^2/g 0.134 cm^3/g	Bulk/Powder, < 74 μm	[122]
FA-GP/Polyethersulfone hybrid	Fly ash	AA $SiO_2/Al_2O_3 = 3.05$	168.3 m^2/g	Bulk/Powder, 150 μm	[123]
FA/Z-GP	Calcinated fly ash	$SiO_2/Al_2O_3 = 1.61^{**}$	–	Bulk/Powder	[124]
FA-GP	Fly ash	AA	131.4 m^2/g	Bulk/Powder, <105 μm	[125]
FA-GP	Coal Fly ash	FM $SiO_2/Al_2O_3 = 1.25^{*}$	93.8 m^2/g 0.62 cm^3/g	Bulk	[8]
FA-GP	Fly ash silica	AA Si/Al = 2.2	31.87 m^2/g 0.12 cm^3/g 15.45 nm	Bulk/Powder, 125–212 μm	[126]
FA-GP	Fly ash F, ≤ 177 μm	AA $SiO_2/Al_2O_3 = 2.97^{**}$	30 m^2/g 0.076 cm^3g	Bulk	[127]
FA/analcime-GP	Fly ash analcime	AA $SiO_2/Al_2O_3 = 2.10^{**}$	–	Bulk/Membrane	[128]
FA-GP	Fly ash	AA $SiO_2/Al_2O_3 = 5.42$	35.97 m^2/g 124 cm^3/kg 9 nm	Bulk/Powder, 150 μm	[31]

GP/AAM	Precursor/additives	Preparation method, prime oxide ratios	Surface Area/Pore Volume/Pore size	Type/form of GP	Ref.
FA-GP	Rice husk ash, waste alum cans	HT SiO_2/Al_2O_3 = 1.82*	36.15 m^2/g 0.097 mL/g 5.4 nm	Bulk	[129]
FA-GP	Fly ash class C and F	AA SiO_2/Al_2O_3 = 6.6** (class C) SiO_2/Al_2O_3 = 10.9** (class F)	2463.64 mm^2/g	–	[130]
FA/MK-GP	bioFA MK 70:30 (w/w)	AA aluminum powder, anionic surfactant	46.3 m^2/g	Foam/Membrane	[131]
IBA-GP/Graphene hybrid	Bottom ash graphene	AA 0.15 wt% Mn^{2+} 19.5 wt% CuO	29.28 m^2/g 0.1078 mL/g 14.77 nm	Bulk/ Particles, 0.180–0.315 mm	[132]
BFS –GP	BFS	AA SiO_2/Al_2O_3 = 4.40*	64.5 m^2/g 0.095 cm^3/g 5.93 nm	Bulk/Powder, 63–125 μm	[32]
BFS –GP	BFS	AA SiO_2/Al_2O_3 = 3.2	–	Bulk/Powder, 63–125 μm	[115]
BFS –GP/graphene hybrid	BFS graphene 0.01 wt%	AA SiO_2/Al_2O_3 = 2.61*	146.17 m^2/g 0.161 mL/g 4.40 nm	Bulk/Powder, 250-315 μm	[133]
BFS –GP/barium modified	BFS	AA SiO_2/Al_2O_3 = 4.00**	63.1 m^2/g 0.070 cm^3/g	Bulk/Powder, 63–125 μm	[134]
BOFS-GP	BOFS	AA SiO_2/Al_2O_3 = 11.5**	30.84 m^2/g 0.091 cm^3/g 11.8 nm	Bulk/Particles, ∼0.1 mm	[39, 40]
Slag-based GP	Slag	SSM SiO_2/Al_2O_3 = 4.02* 0.3 wt% SDS	100.9 m^2/g 7 nm	Porous/Spheres, d ≈ 100 μm	[33]

GP/AAM	Precursor/additives	Preparation method, prime oxide ratios	Surface Area/Pore Volume/Pore size	Type/form of GP	Ref.
Silicomanganese slag-GP	Silicomanganese slag $(NH_4)_6Mo_7O_{24}\cdot 4H_2O$	AA SiO_2/Al_2O_3 = 1.44**	51.79 m^2/g 0.192 mL/g 10.30 nm	Bulk/Particles, 0.16–0.315 mm	[135]
BOFS-GP modified with Ni(II) or Zn(II)	BOFS-GP	AA	Zn/LDS-GP 58.14 m^2/g Ni/LDS-GP 53.42 m^2/g LDS-GP 30.84 m^2/g	Porous/Powder, ~0.1 mm	[41]
Slag-based GP	Slag	SSM SiO_2/Al_2O_3 = 3.08*	87.74 m^2/g	Bulk/Spheres	[34]
Slag-based GP/ Fe_2O_3-hybride	Slag	SSM SiO_2/Al_2O_3 = 70.65** Fe_2O_3/Al_2O_3 = 188 Fe_2O_3/SiO_2 = 2.66	233.8 m^2/g	Bulk/Microspheres, 75–300 μm	[136]
Steel slag/fly ash/analcime-GP	Steel slag fly ash	HT SiO_2/Al_2O_3 = 2.01**	27.25 m^2/g 0.050 cm^3/g 8.12 nm	Bulk	[137]
BFS –GP	BFS	AA SiO_2/Al_2O_3 = 5.26	23.56 m^2/g 73 cm^3/kg 7.8 nm	Bulk/Powder, 150 μm	[31]
EAFS-GP	electric arc furnace slag	AA SiO_2/Al_2O_3 = 2.02*	6.5 m^2/g 0.014 cm^3/g 8.7 nm	Bulk/Powder	[38]
BOFS-GP	Basic Oxygen furnace slag	AA H_2O_2	–	Porous	[43]
Slag-GP/CeO loaded	Slag	SSM SiO_2/Al_2O_3 = 3.31**	186.40 m^2/g 0.352 cm^3/g 7.56 nm	Bulk/Sphere, 75–300 μm	[138]

GP/AAM	Precursor/additives	Preparation method, prime oxide ratios	Surface Area/Pore Volume/Pore size	Type/form of GP	Ref.
Clay-based GP	Kaolin	FM SiO_2/Al_2O_3 = 1.88**	51.3 m^2/g 0.324 cm^3/g 25.25 nm	Bulk/Powder	[58]
clay/gangue microsphere -GP	Kaolin coal gangue 50/50 wt%	AA SiO_2/Al_2O_3 = 4.0	39.74 m2/g 52.00 nm	Bulk/rubbles, 0.45–0.15 mm	[139]
Clay-GP/Fe3O4 hybride	Calcined bentonite clay	AA	2.32 m^2/g 0.008 cm^3/g 13.76 nm	Bulk/Powder	[61]
Clay-GP	Lateritic clay, 58 μm	AA	17.441 m^2/g 0.005 cm^3/g 1,4 nm	Bulk/Powder, 58 μm	[62]
Natural tuff-GP	Volcanic tuff	AA SiO_2/Al_2O_3 = 3.74**	–	Bulk/Powder, < 200 μm	[140]
Alumino silicate-GP	Alumino silicate powder	AA SiO_2/Al_2O_3 = 4	50.1 m^2/g 0.36 cm^3/g 0.04 μm	Bulk/Monoliths or granules	[141]
Synthetic GP	Chemosynthetic Al_2O_3-SiO_2 powder	SSM SiO_2/Al_2O_3 = 2	–	Bulk/Spheres	[142]
Chitosan modified geopolymer	Aluminum salt and silica solution, chitosan	Precipitation SiO_2/Al_2O_3 = 3.06**	–	Bulk/Powder	[143]
OTB-GP	Pyrophyllite mine waste samples	AA SiO_2/Al_2O_3 = 2.39**	–	Bulk/Powder, <45 μm	[55]
OTB-GP	Gold mine waste	FM	74.92 m^2/g	Bulk	[144]
OTB-GP	Gold mine tailings Al_2O_3	FM	74.916 m^2/g	Bulk	[51]
Municipal solid waste-GP	Sludges	FM SiO_2/Al_2O_3 = 3.12**	0.496 m^2/g 9.98 nm	Bulk	[145]

GP/AAM	Precursor/additives	Preparation method, prime oxide ratios	Surface Area/Pore Volume/Pore size	Type/form of GP	Ref.
Municipal solid waste-GP	Municipal solid waste biochar	AA	6.5 m^2/g - -	Bulk	[146]
Dolochar ash based geopolymer	Dolochar < 100 mesh	AA SiO_2/Al_2O_3 = 4.97**	49.91 m^2/g 0.087 cm^3/g 8,9 nm	Bulk/Particles, ≈ 0.1 mm	[147]

Materials: *GP – geopolymer; MK – Metakaolin; FA – fly ash; IBA – incineration bottom ash; BFS – blast furnace slag; BOFS – basic oxygen furnace slag; EAFS – electric arc furnace slag; Z – zeolite; IOT – iron ore tailing; OTB-ore/tailings based;* Procedure: *AA – alkaline activation at moderate temperature; SSM – suspension-solidification method; FM-fusion method; HT- Hydrothermal method;* Additives: *SDS – sodium dodecyl sulfate; CTAB – cetyltrimethylammonium bromide; HDTMABr -hexadecyltrimethylammonium bromide.*

**Calculated using the amounts of raw materials in the slurry.*

***Calculated using the XRF of product.*

Table 1.
GP/AAMs compositions for water and wastewater treatment reported in literature.

depends on the wetting behavior and particle size of the precursors, a good starting point is L/S of 1/3. The production method is easy to upscale that makes feasible large-scale water treatment applications with granular GPs. A more complicated procedure enables the development of granular adsorbents through a suspension and solidification method, resulting in microspheres (<100 μm) [33, 34] or highly porous GPs [93]. Composites of GPs with biopolymers, for instance alginate that possesses the ability to solidification in presence of calcium ions, were also obtained in granulated form [83, 93].

Porous GP/AAM adsorbents contain relatively high volume of voids or pores. The pore sizes usually range from nanometers up to millimeters with the total pore volume ranging from 30 to 90% [153]. Direct foaming, either chemically or mechanically, is seen as the most widely applied foaming approach. The common additives that have been used in the direct foaming methods are hydrogen peroxide [11, 33, 43, 54, 78, 93], Al [131, 154] or Si [155] powders along aside with stabilization agents such surfactants or oils.

Pervious GP/AAM is another promising form of an adsorptive material for water purification. Thus, AAM-based membranes with potential to remove alkaline earth metals [156], and nickel [157] have been reported. Development of porous/pervious GPs/AAMs led to variety of fabricated adsorptive forms such as monoliths, membranes, granules, and self-supported filters. That is opening the versatility of approaches, conventional in water and wastewater treatment practice, yet challenging if the powdered forms used. Separation, regeneration, and surface modification are no longer restricted by the form of production. Porous/pervious materials can be used directly in packed bed adsorbers, and be easily regenerated or retrieved after adsorbent saturation with target substances or contaminants.

3. Properties and performance of GPs/AAMs towards various contaminants

Despite the fact that the first identification of GPs as unconventional construction materials was in 1979 [158], broader applications of GPs/AAMs started in late 90s. Although GPs/AAMs are to be considered by some authors as an economic alternative to zeolites or activated carbons for water purification, the lack of real cases reported is obvious. To urge commercial importance, GP/AAM adsorbents should be readily available, economically feasible, steady in characteristics, and easily regenerated. Several comprehensive reviews on the GP/AAM materials for the water treatment sector have been published just recently [57, 150, 151, 153]. Therefore, in this section the bright and promising works will be highlighted as well as challenges and trends for future studies revealed.

GPs/AAMs for metal(oid)s removal. The adsorption characteristics of individual species and particular conditions of adsorption could be found elsewhere [57, 150, 151, 153]. Here, we would like to emphasize some challenges and gaps, which might be addressed in future studies. There are only several articles discussing selectivity of adsorption on GPs/AAMs alongside the matrix effects. In most of the studies pure mono-element aqueous solutions were implied, and the adsorption characteristics for individual substances- without possible influence of matrix macro-elements have been established. However, GPs/AAMs that are considered as a replacement of zeolites have to demonstrate selectivity under complex matrices in order to have opportunities to promote the implementations in various industrial applications.

In order to obtain adequate adsorption parameters, an excessive alkaline residue in GP/AAM should be washed out properly (pH 7 ± 0.5 within 24 h required) [159].

Otherwise, the increment of pH of aqueous solutions containing heavy metals will favor the hydroxide precipitation process, leading to wrong result interpretation. For porous GPs, washing away the excessive alkalis resulted in the increment of total porosity [11], which led to better performance. Moreover, excessive alkalis were used intentionally to neutralize AMD [42] and remove metal ions. However, a strict protocol must be followed to characterize newly designed materials.

Selective adsorption is relies on several factors such as a metal ion activity, hydration radius and free energy of hydration, and a pore size distribution of GP.

Geopolymerisation by itself could lead to the formation of new ion-exchange sites at the GP surface, but additives in composite formulations could have even higher influence the adsorption characteristics.

An ionic exchange reaction between the heavy metal ions and sodium ions has resulted in heavy metal removal by the metakaolin GP [159]. The adsorption selectivity of heavy metal ions by the GPs at pH 4 in multi-component solution was in the following order: $Pb^{2+} > Cd^{2+} > Cu^{2+} > Cr^{3+}$, while q_e [mg/g]: 100 > 76 > 55 > 10. The order of adsorption was in accordance with the hydrated radius and free energy of hydration for selected ions. However, the free energy of hydration and the activity for Cr^{3+} are all higher compared to those of other metals, though its adsorption rate does not correspond to the assumed order. The selectivity towards Cr^{3+} was be explained through its ionic status. When the pH exceeded 4, Cr^{3+} transforms to $Cr(OH)^{2+}$, which might lead to its lower adsorption ability. It is also noted that at lower pH, the balancing ions present on the GP surface tend to be replaced by the hydrogen ions instead of the metal ions that lead to lower capacity at acidic pH.

Lopez et al. [5] investigated the selectivity of metakaolin-based GPs in multicomponent solutions (Pb^{2+}, Cu^{2+}, Cd^{2+}, Ni^{2+}, Zn^{2+} and Cs^{+}). For a composition with Si/Al ratio 2, the best capacities and selectivity towards Pb^{2+} and Cs^{+} were observed. The adsorption selectivity for the mixture of metal ions was in the following order $Cs^{+} > Pb^{2+} > Cu^{2+} > Zn^{2+} > Ni^{2+} > Cd^{2+}$, while q_m [mg/g]: 43 > 35 > 15 > 3 > 1 > 2. The adsorption capacity for individual elements were higher: 57 mg Pb^{2+}/g > 52 mg Cs^{+}/g > 46 mg Cu^{2+}/g > 14 mg Cd^{2+}/g > 9 mg Zn^{2+}/g > 4 mg Ni^{2+}/g. Moreover, the effect of solution salinity (NaCl, 5% and 10%, wt) was studied, and no considerable effect on the adsorption order of metal ions or GP capacity in multi-composition solution was found. The authors presumed the existence of at least two types of binding sites with different affinities toward the metal ions to explain such a tolerance.

Selectivity of GP composites with zeolite filler was studied by Andrejkovičová et al. [4]. The highest adsorption was observed for Pb^{2+} for all the GPs obtained, while an adsorption order was as follows: $Pb^{2+} > Cd^{2+} > Zn^{2+} > Cu^{2+} > Cr^{3+}$. The adsorption of Cu^{2+} and Cr^{3+} increased as the amount of metakaolin in the GP increased, whereas the composite with 25% zeolite doping had higher adsorption characteristics towards Pb^{2+}, Cd^{2+} and Zn^{2+}. GPs prepared from zeolitic tuff and kaolinitic soil by El-Eswed et al. [160] showed totally different order of adsorption: $Cu^{2+} > Pb^{2+} > Ni^{2+} > Cd^{2+} > Zn^{2+}$. Moreover, the adsorption order strongly depended on the GP composition, although Cu^{2+} and Pb^{2+} adsorption has always prevailed.

The ability of BFS- and metakaolin-based GPs to remove Ni^{2+} and metalloids (As and Sb) in form of oxyanions was shown in [32]. Both adsorbents completely removed Ni^{2+} that most likely was associated with precipitation of its hydroxides on the GPs, while both metalloid oxyanions were adsorbed by BFS-GP equally. Another remarkable merit is that the adsorption capacities were obtained with real matrixes (spiked mine effluents), and were 4.42 mg/g, 0.52 mg/g, and 0.34 mg/g for Ni^{2+}, As^{3+}, and Sb^{3+}, respectively. It is specified by the authors that the low capacities could be a result of competition of some matrix ions (Sr, Ca, Mg, Mn) with the target ions for binding sites.

Researches with increasing frequency pay attention to this problem and try to demonstrate the removal efficiencies with real samples. Removal of Ca^{2+} and Mg^{2+} from intact groundwater was examined in [58] on kaolin-based GP. With adsorbent dose of 1 g/L, the removal rate were 37.5% and 16.2% for Ca^{2+} and Mg^{2+}, respectively. Metakaolin-based GP was tested by Kara et al. [87] for Mn^{2+} and Co^{2+} removal from real wastewater. The removal rates in real wastewater decreased from 97.5% to 53.01% and 94.6% to 39.12% for Co^{2+} and Mn^{2+}, respectively. The results demonstrated that the adsorption performance affected negatively by the coexistence of some other cations and/or anions in the adsorption medium. Bentonite-based GPs were used for heavy metals removal from synthetic wastewater [61]. Porous biomass FA-based GPs were used in [129] for simultaneous removal of heavy metals from wastewater samples. Mixed FA/metakaoline-based GPs were used in [103] for Cu^{2+} removal from real wastewater. In the showcase, the adsorption capacity of GPs towards Cu^{2+} decreased by 27% as compared to synthetic samples. Sithole et al. treated acidic industrial effluents by FA/BOFS-based GPs [42, 43]. New GPs containing hollow gangue microsphere were applied for Zn^{2+} removal from smelting plant wastewater in [93]. At an adsorbent dose of 30 g/L, a complete Zn removal was observed. The distinctive aspect of the reported cases was that a complex composition of treated solutions is likely to decrease substantially capacity of the GP. Thus, the adsorption capacities obtained for the ideal laboratory conditions should be primary used as the guiding not decision-making parameters.

GPs/AAMs for removal of other inorganic ions. Besides metal(oid)s, GPs/AAMs were examined for removal of ammonium and various anions. Luukkonen et al. [149, 152, 161] showed potential of metakaolin-based GPs to remove ammonium. The optimized GP composition was proposed and manufactured in both powder and granular forms. The efficiency of removal was demonstrated in municipal wastewaters (primary and secondary effluents) as well as landfill leachates. Metakaolin-based GPs prepared from commercial and waste metakaolin were able to effectively remove ammonium from synthetic and wastewater samples [59]. In fact, GPs prepared from paper mill fiber sludge showed better selectivity in the presence of competing ions under real matrix conditions. Bai and Colombo prepared metakaolin-based GP foams in the form of monolithic porous filters [162, 163]. The filter was able to remove up to 95.3% of ammonium from runoff waters at the initial concentration of 3 mg/L.

The removal of phosphorus was attempted in [10] with a pervious FA-based GP. The removal rate increased with the increase of pH. Up to 85% of phosphorus were removed from a treated wastewater. Simultaneous removal of ammonium and phosphate by composite metakaolin/BFS-based GPs was demonstrated in [91]. Phosphate removal was enhanced in presence of ammonium. At slightly alkaline conditions (pH 7–8), the removal rate towards phosphate ions was relatively high (>86%), whereas the ammonium removal up to 35% was also achieved. FA-, BFS- and fiber sludge GPs were investigated as promising adsorbents for phosphorous removal from diluted solutions. The capacities at initial phosphate concentration of 100 mg/L are 26 mg PO4/g for BFS-GP, 36 mg PO4/g for FAF-GP, and 43 mg PO4/g for FSHCa-GP [115].

Sulfate ions were removed by barium-modified BFS-based GPs [134]. Adsorption capacities were 91.1 and 119.0 mg SO_4/g for model solution and mine effluent, respectively. The surface complexation or precipitation of barium sulfate were suggested as probable removal mechanisms.

Removal of halides by GPs/AAMs is an emerging topic. For this end, composite or functionalized materials are designed. Removal of F^- was demonstrated by slag-based GP microspheres modified with CeO [138], Fe_2O_3 [136], and bivalent metallic species [41] with capacities towards the contaminant 127.7 mg/g, 59.8 mg/g, and 60 mg/g (zinc impregnated BOFS-GP), respectively. A metakaolin-based GP functionalized by surfactant was developed for efficient removal of radioactive iodide [97]. High concentrations of competitive anions had limited influence on the adsorption process.

GPs/AAMs for removal of organic substances. In fact, GPs contained residual metal oxides could have potential catalytic performance. Thus, GPs based on industrial by-products such as FA, BFS, or their mixtures with silica fume and aloxite demonstrated catalytic activity under visible light irradiation. The descriptive list of organic substances removed by GPs could be found in reviews [57, 150, 151, 153]. Manly, cationic and neutral dyes were investigated as targets in recent studies, although removal of fecal coliforms [10], volatile organic compound [77], and tetracycline [164] was reported.

Oxidative degradation or photodegradation after adsorption have been specified by authors as primary mechanisms of organic pollutants' removal. Although conventional GPs have been reported for these purposes [86, 89, 104, 126, 139], they would rather have had low adsorption/degradation characteristics. Hybrid or composite materials were proposed to improve the removal efficiency of organic pollutants. Thus, graphene [120, 132, 133, 165], TiO2 [88, 98, 105], CdS [142], various metal oxides [101, 106, 135] were introduced in GP matrix in order to enhance degradation abilities of resulting materials.

4. Regeneration of GPs/AAMs and further resource recovery options

In last a few decades, significant improvements were made in both efficiency and economy in removal of metal(oid)s and other substances by adsorbents. Nevertheless, regeneration and recycling of used adsorbents, or recovery of the removed species from the desorbing agents are still rarely reported. For regeneration and reuse of GPs/AAMs, various possible regenerating agents such as acids, alkalis and chelating agents could be used. Only a few of the reported studies were focused on recovery of adsorbed (from saturated adsorbents) and desorbed (from regenerating agents) metals [11, 87, 96, 131]. However, for industrial application and success completion of new GP/AAM adsorbents on the market, research studies on number of adsorption–desorption cycles are in high demand. Moreover, revenues gathered from resource recovery options will have a decisive role in further technology implementation.

The regeneration of metakaolin-based GP by sodium chloride under alkaline conditions after ammonium adsorption for the first time were demonstrated in [152]. Three adsorption–desorption cycles were carried out with a steady removal efficiency. Sodium chloride and sulfate, potassium sulfate and phosphate were studied in [59] as regenerating agents for saturated metakaolin-based GPs. Sodium sulfate showed better results during five cycles under continuous sorption–desorption experiment, only 34% of an initial overall capacity of the GP were lost. Sodium chloride regenerant was also efficient, but only 55% of ammonium could be removed after 5th desorption cycle. The same adsorbents were used to test a nitrogen recovery option in a laboratory-scale demonstration setup [166]. The layout consisted of an adsorption/desorption unit and Liqui-Cel® membrane. A liquid phase obtained during adsorbent regeneration was purified in the membrane contactor in order to recover ammonium nitrogen as ammonium sulfate or phosphate. The purified regeneration solution was used repeatedly for further adsorbent regeneration. Several regeneration-purification cycles were conducted to estimate system sustainability and chemical consumption demand. Operational conditions of a membrane process such as shellside and lumenside feed flows, temperature, and pH were adjusted to gain maximal capacity of the setup. One membrane contactor (2.5 $\times$ 8-inch Liqui-Cel) was used under following operational conditions: 100 L/h shellside and 60 L/h lumenside feed flows, 40°C working temperature, pH $\geq$ 10.

Technical sulfuric or phosphoric acids, up to 5%, were used as lumenside phases. The concentration of ammonium-content salt in a resulting received phase were 17% and 22% for phosphate and sulfate salt, respectively.

Metal recovery from GPs/AMMs via ion-exchange mechanism can only take place if physical adsorption occurred and the pH was low enough to prevent precipitation of metal hydroxide during adsorption process. Acids of over 0.1 M strength affect the structure of the GPs, and while metals are regenerated by acid washing, the reuse of adsorbents are diminished both in batch [11] as also in continuous mode [87, 167] experiments. Mild acid washing with 0.01 M H_2SO_4 or HNO_3 removed metals from GPs efficiently in short time (1–2 h). It has also been shown that the adsorption capacity after mild acid washing could increase [131], which could be explained by exchange of Na^+ with easier replaceable H^+ cations. Selective desorption of copper has been observed by ammonia. A linear desorption ability with respect to ammonia concentration was observed, and complete desorption being possible by 10% ammonia solution [50, 61].

Sequential desorption tests of Cd^{2+} have been conducted on a loaded metakaolin GP, establishing the percentages of physically adsorbed, ion-exchangeable, EDTA extractable, and residual forms of metal [96]. The authors showed that physical adsorption is negligible, and ion-exchange with $MgCl_2$ constituted to only 2–8% of adsorbed Cd^{2+}. The bulk amount of Cd^{2+} adsorbed by the metakaolin GP was EDTA extractable, and the adsorbent remained 85% of its adsorption capacity after EDTA desorption for 5 cycles. Luukkonen [32] and Naghsh [58] suggested the efficient metal desorption by 5% NaCl. However, care must be taken since the balancing ions can form a positively charged film on the adsorbent surfaces. El Esweed et al. have achieved ion-exchange based desorption of Cu^{2+} by 0.1 M NaCl [160]. From all the studies reported, only Cd^{2+} has been shown to be desorbed at pH > 8 with NaOH solution, achieving 24–84% desorption [64].

5. Environmental impact and costs of treatment with GPs/AAMs

An efficient use of GPs/AAMs in real wastewater treatment practices including economic evaluation is little investigated. Above all, these adsorbents show rather low selectivity, and therefore the ubiquitous metal ions (Na^+, Ca^{2+}, Mg^{2+}, Fe^{3+}) present in wastewater solutions demonstrate either competing interaction with the target ions, or the interaction has not been studied [57]. Additionally, for economic and ecological assessment is essential that the adsorbent would be regenerable [168]. To be economically successful, exhausted adsorbents need to pass the non-hazardous leaching criteria of the adsorbed materials, while the amount of waste regenerated should be as little as possible. This means that the adsorption-regeneration cycle needs to be performed as often as possible. And yet, afterwards the adsorbent needs to find end storage place, e.g. in tailing pond, or further use, e.g. as binder, filler, or soil amendment.

Adsorption capacity of a powdered GP is usually higher, but technical implementation of powdered forms requires precise dosing, contact vessel with stirring, solid–liquid separation step, and transfer of exhausted adsorbent to regeneration vessel. The powder can then be regenerated by addition of suitable regenerant, e.g. mild acid, separated, and dried prior to the next adsorption cycle.

Technically, the use of granular forms is an easier option. However, the size of the column vs. wastewater stream can easily become very large, as granules per se, are larger particles and adsorption is a surface process. This puts additional burden on geopolymer production as the overall capacity should be sufficient, and the granules will need to show suitable compressive strength to withstand the gravimetric pressure in the purification column. Conversely, regeneration is technically easily realized by counter flow of regeneration liquid through the column.

Economic evaluation therefore needs to take these considerations into account during CAPEX estimation. OPEX, in turn, is not only the ongoing replacement of exhausted adsorbent, electricity consumed, maintenance, staff, and regeneration chemicals, but also the transportation costs of adsorbents, which can be high at low adsorption capacity.

As a thought experiment, an example of 55 mg/g adsorption capacity of copper adsorbent, with 85% cycling capacity has a 47 mg/g adsorption capacity after desorption cycle, shall be considered. For a mine effluent or process water with 5 mg/L Cu^{2+} and a flow of 200 m^3/h requires about 21 kg adsorbent per hour. The price of GPs is given as 1–1.5 € per kg [103], and as such the treatment costs of merely 1 h would be between 10 and 21 €. Regeneration up to 20 times gives more realistic cost factors, of 0.5–1 € per h, only for adsorbent costs. It becomes quickly clear that without regeneration, high efficiency, and selectivity GPs/AAMs will be too expensive for wastewater treatment.

6. Conclusions

Much work has been done on the adsorption properties of GPs/AAMs towards a wide variety of inorganic pollutants during the last decade. While the effect of competing ions in real water samples remain an issue, the incorporation of new composite materials and the tailoring of reaction conditions have a high potential to increase their selectivity as adsorbents. However, more and more authors have understood the need to regenerate adsorbents and research are being conducted on the recovery of valuable materials, such as metals or nutrients. The recovery of high energy products from side streams utilizing adsorbents made from industrial side streams, will bring circular economy towards the next level. It is also of interest, to cover the costs of water treatment by the revenue of removed materials. While still much work needs to be done, the authors remain confident, that GPs/AAMs will continue to have a prominent place in wastewater treatment.

Acknowledgements

The research was partially funded by European Regional Development Fund (Leverage from the EU, WaterPro project № A74635; Keski-Pohjanmaan liitto/Kainuun liitto/Pohjois-Pohjanmaan liitto) and by Maa- ja vesitekniikan tuki (№ 13-8271-17).

Author details

Tatiana Samarina*, Esther Takaluoma and Outi Laatikainen
KAMK Kajaani University of Applied Sciences, Kajaani, Finland

*Address all correspondence to: tatiana.samarina@kamk.fi

References

[1] Bennich T, Weitz N, Carlsen H. Deciphering the scientific literature on SDG interactions: A review and reading guide. Science of The Total Environment 2020;728:138405. DOI: 10.1016/j.scitotenv.2020.138405.

[2] World Bank Group. The Growing Role of Minerals and Metals for a Low Carbon Future. World Bank, Washington, DC; 2017. DOI: 10.1596/28312.

[3] Takaluoma E, Samarina T. Valorisation of Sidestream from lithium Refining - Activation of Analcime. Proceedings of the 6th World Congress on New Technologies (NewTech'20), Virtual: AVESTIA; 2020. DOI: 10.11159/icepr20.149.

[4] Andrejkovičová S, Sudagar A, Rocha J, Patinha C, Hajjaji W, da Silva EF, et al. The effect of natural zeolite on microstructure, mechanical and heavy metals adsorption properties of metakaolin based geopolymers. Applied Clay Science 2016;126:141–152. DOI: 10.1016/j.clay.2016.03.009.

[5] López FJ, Sugita S, Tagaya M, Kobayashi T. Metakaolin-Based Geopolymers for Targeted Adsorbents to Heavy Metal Ion Separation. Journal of Materials Science and Chemical Engineering 2014;2:16–27. DOI: 10.4236/msce.2014.27002.

[6] Ariffin N, Abdullah MMAB, Zainol MRRMA, Baltatu MS, Jamaludin L. Effect of Solid to Liquid Ratio on Heavy Metal Removal by Geopolymer-Based Adsorbent. IOP Conf Ser: Mater Sci Eng 2018;374:012045. DOI: 10.1088/1757-899X/374/1/012045.

[7] Gollakota ARK, Volli V, Shu C-M. Progressive utilisation prospects of coal fly ash: A review. Sci Total Environ 2019;672:951–989. DOI: 10.1016/j.scitotenv.2019.03.337.

[8] Tian Q, Sasaki K. Structural characterizations of fly ash-based geopolymer after adsorption of various metal ions. Environmental Technology 2019;0:1–11. DOI: 10.1080/09593330.2019.1649469.

[9] Mužek MN, Svilović S, Ugrina M, Zelić J. Removal of copper and cobalt ions by fly ash-based geopolymer from solutions-equilibrium study. Desalination and Water Treatment 2016;57:10689–10699. DOI: 10.1080/19443994.2015.1040077.

[10] Jo M, Soto L, Arocho M, St John J, Hwang S. Optimum mix design of fly ash geopolymer paste and its use in pervious concrete for removal of fecal coliforms and phosphorus in water. Construction and Building Materials 2015;93:1097–1104. DOI: 10.1016/j.conbuildmat.2015.05.034.

[11] Novais RM, Buruberri LH, Seabra MP, Labrincha JA. Novel porous fly-ash containing geopolymer monoliths for lead adsorption from wastewaters. J Hazard Mater 2016;318:631–640. DOI: 10.1016/j.jhazmat.2016.07.059.

[12] Yao Z. Fly Ash: Sources, Applications and Potential Environmental Impacts. 2014.

[13] De Gisi S, Lofrano G, Grassi M, Notarnicola M. Characteristics and adsorption capacities of low-cost sorbents for wastewater treatment: A review. Sustainable Materials and Technologies 2016;9:10–40. DOI: 10.1016/j.susmat.2016.06.002.

[14] Ahmaruzzaman M. A review on the utilization of fly ash. Progress in Energy and Combustion Science 2010;36:327–363. DOI: 10.1016/j.pecs.2009.11.003.

[15] Ji Z, Pei Y. Geopolymers produced from drinking water treatment residue

and bottom ash for the immobilization of heavy metals. Chemosphere 2019;225: 579–587. DOI: 10.1016/j.chemosphere.2019.03.056.

[16] Ji Z, Pei Y. Bibliographic and visualized analysis of geopolymer research and its application in heavy metal immobilization: A review. Journal of Environmental Management 2019; 231:256–267. DOI: 10.1016/j.jenvman.2018.10.041.

[17] Maldonado-Alameda À, Giro-Paloma J, Alfocea-Roig A, Formosa J, Chimenos JM. Municipal Solid Waste Incineration Bottom Ash as Sole Precursor in the Alkali-Activated Binder Formulation. Applied Sciences 2020;10: 4129. DOI: 10.3390/app10124129.

[18] Al-Ghouti MA, Khan M, Nasser MS, Saad KA, Heng OE. Application of geopolymers synthesized from incinerated municipal solid waste ashes for the removal of cationic dye from water. PLOS ONE 2020;15:e0239095. DOI: 10.1371/journal.pone.0239095.

[19] Silva RV, de Brito J, Lynn CJ, Dhir RK. Use of municipal solid waste incineration bottom ashes in alkali-activated materials, ceramics and granular applications: A review. Waste Management 2017;68:207–220. DOI: 10.1016/j.wasman.2017.06.043.

[20] Chiang YW, Ghyselbrecht K, Santos RM, Meesschaert B, Martens JA. Synthesis of zeolitic-type adsorbent material from municipal solid waste incinerator bottom ash and its application in heavy metal adsorption. Catalysis Today 2012;190:23–30. DOI: 10.1016/j.cattod.2011.11.002.

[21] Hong S-H, Shin M-C, Lee J, Lee C-G, Song D-S, Um B-H, et al. Recycling of bottom ash derived from combustion of cattle manure and its adsorption behaviors for Cd(II), Cu(II), Pb(II), and Ni(II). Environ Sci Pollut Res 2020. DOI: 10.1007/s11356-020-11719-7.

[22] Zhang HY, Zheng Y, Hu HT, Qi JY. Use of Municipal Solid Waste Incineration Bottom Ash in Adsorption of Heavy Metals. Key Engineering Materials 2011;474–476:1099–1102. DOI: 10.4028/www.scientific.net/KEM.474-476.1099.

[23] Joseph A, Snellings R, Van den Heede P, Matthys S, De Belie N. The Use of Municipal Solid Waste Incineration Ash in Various Building Materials: A Belgian Point of View. Materials 2018; 11:141. DOI: 10.3390/ma11010141.

[24] Zhu W, Rao XH, Liu Y, Yang E-H. Lightweight aerated metakaolin-based geopolymer incorporating municipal solid waste incineration bottom ash as gas-forming agent. Journal of Cleaner Production 2018;177:775–781. DOI: 10.1016/j.jclepro.2017.12.267.

[25] Fabricius A-L, Renner M, Voss M, Funk M, Perfoll A, Gehring F, et al. Municipal waste incineration fly ashes: from a multi-element approach to market potential evaluation. Environmental Sciences Europe 2020;32: 88. DOI: 10.1186/s12302-020-00365-y.

[26] Quina MJ, Bontempi E, Bogush A, Schlumberger S, Weibel G, Braga R, et al. Technologies for the management of MSW incineration ashes from gas cleaning: New perspectives on recovery of secondary raw materials and circular economy. Science of The Total Environment 2018;635:526–542. DOI: 10.1016/j.scitotenv.2018.04.150.

[27] Zhuang XY, Chen L, Komarneni S, Zhou CH, Tong DS, Yang HM, et al. Fly ash-based geopolymer: clean production, properties and applications. Journal of Cleaner Production 2016;125: 253–267. DOI: 10.1016/j.jclepro.2016.03.019.

[28] Wang GC. 11 - Slag use as an aggregate in concrete and cement-based materials. In: Wang GC, editor. The Utilization of Slag in Civil Infrastructure

Construction, Woodhead Publishing; 2016, p. 239–74. DOI: 10.1016/B978-0-08-100381-7.00011-2.

[29] Siddique R, Iqbal Khan M. Ground Granulated Blast Furnace Slag. In: Siddique R, Khan MI, editors. Supplementary Cementing Materials, Berlin, Heidelberg: Springer; 2011, p. 121–173. DOI: 10.1007/978-3-642-17866-5_3.

[30] Gupta VK, Carrott PJM, Carrott MMLR, Suhas. Low-Cost Adsorbents: Growing Approach to Wastewater Treatment—a Review. Critical Reviews in Environmental Science and Technology 2009;39:783–842. DOI: 10.1080/10643380801977610.

[31] Medina TJ, Arredondo SP, Corral R, Jacobo A, Zarraga RA, Rosas CA, et al. Microstructure and Pb(2+)Adsorption Properties of Blast Furnace Slag and Fly Ash based Geopolymers. Minerals 2020;10:808. DOI: 10.3390/min10090808.

[32] Luukkonen T, Runtti H, Niskanen M, Tolonen E-T, Sarkkinen M, Kemppainen K, et al. Simultaneous removal of Ni(II), As(III), and Sb(III) from spiked mine effluent with metakaolin and blast-furnace-slag geopolymers. Journal of Environmental Management 2016;166:579–588. DOI: 10.1016/j.jenvman.2015.11.007.

[33] Tang Q, Wang K, Yaseen M, Tong Z, Cui X. Synthesis of highly efficient porous inorganic polymer microspheres for the adsorptive removal of Pb2+ from wastewater. J Clean Prod 2018;193:351–362. DOI: 10.1016/j.jclepro.2018.05.094.

[34] Tang Q, Wang K, Su J, Shen Y, Yang S, Ge Y, et al. Facile fabrication of inorganic polymer microspheres as adsorbents for removing heavy metal ions. Mater Res Bull 2019;113:202–208. DOI: 10.1016/j.materresbull.2019.02.009.

[35] Khalid HR, Lee NK, Park SM, Abbas N, Lee HK. Synthesis of geopolymer-supported zeolites via robust one-step method and their adsorption potential. Journal of Hazardous Materials 2018;353:522–533. DOI: 10.1016/j.jhazmat.2018.04.049.

[36] Manchisi J, Matinde E, Rowson NA, Simmons MJH, Simate GS, Ndlovu S, et al. Ironmaking and Steelmaking Slags as Sustainable Adsorbents for Industrial Effluents and Wastewater Treatment: A Critical Review of Properties, Performance, Challenges and Opportunities. Sustainability 2020;12:2118. DOI: 10.3390/su12052118.

[37] Naidu TS, Sheridan CM, van Dyk LD. Basic oxygen furnace slag: Review of current and potential uses. Minerals Engineering 2020;149:106234. DOI: 10.1016/j.mineng.2020.106234.

[38] Nikolić I, Đurović D, Tadić M, Radmilović VV, Radmilović VR. Adsorption kinetics, equilibrium, and thermodynamics of Cu2+ on pristine and alkali activated steel slag. Chemical Engineering Communications 2020;207:1278–1297. DOI: 10.1080/00986445.2019.1685986.

[39] Sarkar C, Basu JK, Samanta AN. Removal of Ni2+ ion from waste water by Geopolymeric Adsorbent derived from LD Slag. Journal of Water Process Engineering 2017;17:237–244. DOI: 10.1016/j.jwpe.2017.04.012.

[40] Sarkar C, Basu JK, Samanta AN. Synthesis of mesoporous geopolymeric powder from LD slag as superior adsorbent for Zinc (II) removal. Advanced Powder Technology 2018;29:1142–1152. DOI: 10.1016/j.apt.2018.02.005.

[41] Sarkar C, Basu JK, Samanta AN. Experimental and kinetic study of fluoride adsorption by Ni and Zn modified LD slag based geopolymer. Chemical Engineering Research and

Design 2019;142:165–175. DOI: 10.1016/j.cherd.2018.12.006.

[42] Sithole NT, Ntuli F, Okonta F. Fixed bed column studies for decontamination of acidic mineral effluent using porous fly ash-basic oxygen furnace slag based geopolymers. Miner Eng 2020;154: 106397. DOI: 10.1016/j.mineng.2020.106397.

[43] Sithole NT, Ntuli F, Okonta F. Synthesis and evaluation of basic oxygen furnace slag based geopolymers for removal of metals and sulphates from acidic industrial effluent-column study. J Water Process Eng 2020;37: 101518. DOI: 10.1016/j.jwpe.2020.101518.

[44] Alhassan SI, He Y, Huang L, Wu B, Yan L, Deng H, et al. A review on fluoride adsorption using modified bauxite: Surface modification and sorption mechanisms perspectives. Journal of Environmental Chemical Engineering 2020;8:104532. DOI: 10.1016/j.jece.2020.104532.

[45] Alshaebi F, Yaacob WZ, Samsudin A, Esmail A. Arsenic Adsorption on Bauxite Mineral Using Batch Equilibrium Test. American Journal of Applied Sciences 2009;6. DOI: 10.3844/ajassp.2009.1826.1830.

[46] Hertel T, Pontikes Y. Geopolymers, inorganic polymers, alkali-activated materials and hybrid binders from bauxite residue (red mud) – Putting things in perspective. Journal of Cleaner Production 2020;258:120610. DOI: 10.1016/j.jclepro.2020.120610.

[47] Norouzi S, Badii K, Doulati Ardejani F. Activated bauxite waste as an adsorbent for removal of Acid Blue 92 from aqueous solutions. Water Sci Technol 2010;62:2491–2500. DOI: 10.2166/wst.2010.514.

[48] Taneez M, Hurel C, Marmier N, Lefèvre G. Adsorption of inorganic pollutants on bauxite residues: An example of methodology to simulate adsorption in complex solids mixtures. Applied Geochemistry 2017;78:272–278. DOI: 10.1016/j.apgeochem.2017.01.007.

[49] Ren B, Zhao Y, Bai H, Kang S, Zhang T, Song S. Eco-friendly geopolymer prepared from solid wastes: A critical review. Chemosphere 2020:128900. DOI: 10.1016/j.chemosphere.2020.128900.

[50] Khalid MK, Agarwal V, Wilson BP, Takaluoma E, Lundstrom M. Applicability of solid process residues as sorbents for the treatment of industrial wastewaters. J Clean Prod 2020;246: 118951. DOI: 10.1016/j.jclepro.2019.118951.

[51] Demir F, Derun EM. Modelling and optimization of gold mine tailings based geopolymer by using response surface method and its application in Pb2+ removal. J Clean Prod 2019;237:117766. DOI: 10.1016/j.jclepro.2019.117766.

[52] Tian Q, Nakama S, Sasaki K. Immobilization of cesium in fly ash-silica fume based geopolymers with different Si/Al molar ratios. Sci Total Environ 2019;687:1127–1137. DOI: 10.1016/j.scitotenv.2019.06.095.

[53] Shakouri S, Bayer O, Erdogan ST. Development of silica fume-based geopolymer foams. Constr Build Mater 2020;260:120442. DOI: 10.1016/j.conbuildmat.2020.120442.

[54] Duan P, Yan C, Zhou W, Ren D. Development of fly ash and iron ore tailing based porous geopolymer for removal of Cu(II) from wastewater. Ceram Int 2016;42:13507–13518. DOI: 10.1016/j.ceramint.2016.05.143.

[55] Panda L, Rath SS, Rao DS, Nayak BB, Das B, Misra PK. Thorough understanding of the kinetics and mechanism of heavy metal adsorption onto a pyrophyllite mine waste based geopolymer. Journal of Molecular

Liquids 2018;263:428–441. DOI: 10.1016/j.molliq.2018.05.016.

[56] Lemougna PN, Wang K, Tang Q, Melo UC, Cui X. Recent developments on inorganic polymers synthesis and applications. Ceramics International 2016;42:15142–15159. DOI: 10.1016/j.ceramint.2016.07.027.

[57] Rasaki SA, Bingxue Z, Guarecuco R, Thomas T, Minghui Y. Geopolymer for use in heavy metals adsorption, and advanced oxidative processes: A critical review. Journal of Cleaner Production 2019;213:42–58. DOI: 10.1016/j.jclepro.2018.12.145.

[58] Naghsh M, Shams K. Synthesis of a kaolin-based geopolymer using a novel fusion method and its application in effective water softening. Applied Clay Science 2017;146:238–245. DOI: 10.1016/j.clay.2017.06.008.

[59] Samarina T, Takaluoma E. Metakaolin-Based Geopolymers for Removal of Ammonium from Municipal Wastewater, 2019. DOI: 10.11159/icepr19.195.

[60] Khalifa AZ, Cizer Ö, Pontikes Y, Heath A, Patureau P, Bernal SA, et al. Advances in alkali-activation of clay minerals. Cement and Concrete Research 2020;132:106050. DOI: 10.1016/j.cemconres.2020.106050.

[61] Maleki A, Hajizadeh Z, Sharifi V, Emdadi Z. A green, porous and eco-friendly magnetic geopolymer adsorbent for heavy metals removal from aqueous solutions. J Clean Prod 2019;215:1233–1245. DOI: 10.1016/j.jclepro.2019.01.084.

[62] Ghani U, Hussain S, Noor-ul-Amin, Imtiaz M, Ali Khan S. Laterite clay-based geopolymer as a potential adsorbent for the heavy metals removal from aqueous solutions. Journal of Saudi Chemical Society 2020;24:874–884. DOI: 10.1016/j.jscs.2020.09.004.

[63] Hayashi S, Onuma T, Kagaya F. Fixing of electrophoretically deposited natural zeolite particles by geopolymer reaction for heavy metal ion adsorbents. Journal of the Ceramic Society of Japan 2017;125:894–898. DOI: 10.2109/jcersj2.17134.

[64] Javadian H, Ghorbani F, Tayebi H, Asl SH. Study of the adsorption of Cd (II) from aqueous solution using zeolite-based geopolymer, synthesized from coal fly ash; kinetic, isotherm and thermodynamic studies. Arabian Journal of Chemistry 2015;8:837–849. DOI: 10.1016/j.arabjc.2013.02.018.

[65] Wang S, Peng Y. Natural zeolites as effective adsorbents in water and wastewater treatment. Chemical Engineering Journal 2010;156:11–24. DOI: 10.1016/j.cej.2009.10.029.

[66] Li Y, Li L, Yu J. Applications of Zeolites in Sustainable Chemistry. Chem 2017;3:928–949. DOI: 10.1016/j.chempr.2017.10.009.

[67] Delkash M, Ebrazi Bakhshayesh B, Kazemian H. Using zeolitic adsorbents to cleanup special wastewater streams: A review. Microporous and Mesoporous Materials 2015;214:224–241. DOI: 10.1016/j.micromeso.2015.04.039.

[68] Yilmaz B, Müller U. Catalytic Applications of Zeolites in Chemical Industry. Top Catal 2009;52:888–895. DOI: 10.1007/s11244-009-9226-0.

[69] Montalvo S, Huiliñir C, Borja R, Sánchez E, Herrmann C. Application of zeolites for biological treatment processes of solid wastes and wastewaters – A review. Bioresource Technology 2020;301:122808. DOI: 10.1016/j.biortech.2020.122808.

[70] Khaleque A, Alam MM, Hoque M, Mondal S, Haider JB, Xu B, et al. Zeolite synthesis from low-cost materials and environmental applications: A review. Environmental Advances 2020;2:

100019. DOI: 10.1016/j.envadv.2020.100019.

[71] Rożek P, Król M, Mozgawa W. Geopolymer-zeolite composites: A review. Journal of Cleaner Production 2019;230:557–579. DOI: 10.1016/j.jclepro.2019.05.152.

[72] Alshaaer M, Zaharaki D, Komnitsas K. Microstructural characteristics and adsorption potential of a zeolitic tuff–metakaolin geopolymer. Desalination and Water Treatment 2015;56:338–345. DOI: 10.1080/19443994.2014.938306.

[73] Deng L, Xu Q, Wu H. Synthesis of Zeolite-like Material by Hydrothermal and Fusion Methods Using Municipal Solid Waste Fly Ash. Procedia Environmental Sciences 2016;31: 662–667. DOI: 10.1016/j.proenv.2016.02.122.

[74] Visa M. Synthesis and characterization of new zeolite materials obtained from fly ash for heavy metals removal in advanced wastewater treatment. Powder Technology 2016; 294:338–347. DOI: 10.1016/j.powtec.2016.02.019.

[75] Rios C, Williams C, Castellanos OM. A comparative study on the use of zeolite-like materials in the removal of heavy metals and ammonium from wastewaters. RP 2020;18. DOI: 10.15665/rp.v18i2.2187.

[76] Toniolo N, Boccaccini AR. Fly ash-based geopolymers containing added silicate waste. A review. Ceramics International 2017;43:14545–14551. DOI: 10.1016/j.ceramint.2017.07.221.

[77] Gasca-Tirado JR, Manzano-Ramírez A, Vazquez-Landaverde PA, Herrera-Díaz EI, Rodríguez-Ugarte ME, Rubio-Ávalos JC, et al. Ion-exchanged geopolymer for photocatalytic degradation of a volatile organic compound. Materials Letters 2014;134: 222–224. DOI: 10.1016/j.matlet.2014.07.090.

[78] Ge Y, Cui X, Kong Y, Li Z, He Y, Zhou Q. Porous geopolymeric spheres for removal of Cu(II) from aqueous solution: Synthesis and evaluation. Journal of Hazardous Materials 2015; 283:244–251. DOI: 10.1016/j.jhazmat.2014.09.038.

[79] Zhang M, Zhao M, Zhang G, Nowak P, Coen A, Tao M. Calcium-free geopolymer as a stabilizer for sulfate-rich soils. Applied Clay Science 2015; 108:199–207. DOI: 10.1016/j.clay.2015.02.029.

[80] Duan J, Li J, Lu Z. One-step facile synthesis of bulk zeolite A through metakaolin-based geopolymer gels. J Porous Mater 2015;22:1519–1526. DOI: 10.1007/s10934-015-0034-6.

[81] Tang Q, Ge Y, Wang K, He Y, Cui X. Preparation and characterization of porous metakaolin-based inorganic polymer spheres as an adsorbent. Materials & Design 2015;88:1244–1249. DOI: 10.1016/j.matdes.2015.09.126.

[82] Kovářík T, Křenek T, Pola M, Rieger D, Kadlec J, Franče P. Ceramic-like open-celled geopolymer foam as a porous substrate for water treatment catalyst. IOP Conf Ser: Mater Sci Eng 2017;175:012044. DOI: 10.1088/1757-899X/175/1/012044.

[83] Ge Y, Cui X, Liao C, Li Z. Facile fabrication of green geopolymer/alginate hybrid spheres for efficient removal of Cu(II) in water: Batch and column studies. Chemical Engineering Journal 2017;311:126–134. DOI: 10.1016/j.cej.2016.11.079.

[84] Kara İ, Yilmazer D, Akar ST. Metakaolin based geopolymer as an effective adsorbent for adsorption of zinc(II) and nickel(II) ions from aqueous solutions. Applied Clay Science

2017;139:54–63. DOI: 10.1016/j.clay.2017.01.008.

[85] Singhal A, Gangwar BP, Gayathry JM. CTAB modified large surface area nanoporous geopolymer with high adsorption capacity for copper ion removal. Applied Clay Science 2017;150:106–114. DOI: 10.1016/j.clay.2017.09.013.

[86] Novais RM, Ascensão G, Tobaldi DM, Seabra MP, Labrincha JA. Biomass fly ash geopolymer monoliths for effective methylene blue removal from wastewaters. Journal of Cleaner Production 2018;171:783–794. DOI: 10.1016/j.jclepro.2017.10.078.

[87] Kara I, Tunc D, Sayin F, Akar ST. Study on the performance of metakaolin based geopolymer for Mn(II) and Co(II) removal. Applied Clay Science 2018;161: 184–193. DOI: 10.1016/j.clay.2018.04.027.

[88] Bravo PI, Malenab RA, Shimizu E, Yu DE, Promentilla MA. Synthesis of geopolymer spheres with photocatalytic activity. MATEC Web Conf 2019;268: 04007. DOI: 10.1051/matecconf/201926804007.

[89] Novais RM, Carvalheiras J, Tobaldi DM, Seabra MP, Pullar RC, Labrincha JA. Synthesis of porous biomass fly ash-based geopolymer spheres for efficient removal of methylene blue from wastewaters. Journal of Cleaner Production 2019;207: 350–362. DOI: 10.1016/j.jclepro.2018.09.265.

[90] Yan C, Guo L, Ren D, Duan P. Novel composites based on geopolymer for removal of Pb(II). Materials Letters 2019;239:192–195. DOI: 10.1016/j.matlet.2018.12.105.

[91] Samarina T, Takaluoma E. Simultaneous Removal of Nutrients by Geopolymers Made From Industrial By-Products. Proceedings of the 5th World Congress on New Technologies (NewTech'19), Lisbone: AVESTIA; 2019. DOI: 10.11159/ICEPR19.169.

[92] El Alouani M, Alehyen S, El Achouri M, Taibi M. Preparation, Characterization, and Application of Metakaolin-Based Geopolymer for Removal of Methylene Blue from Aqueous Solution. Journal of Chemistry 2019;2019:e4212901. DOI: https://doi.org/10.1155/2019/4212901.

[93] Yan S, Zhang F, Wang L, Rong Y, He P, Jia D, et al. A green and low-cost hollow gangue microsphere/geopolymer adsorbent for the effective removal of heavy metals from wastewaters. Journal of Environmental Management 2019; 246:174–183. DOI: 10.1016/j.jenvman.2019.05.120.

[94] Chen X, Niu Z, Zhang H, Guo Y, Liu M, Zhou M. Study on the metakaolin-based geopolymer pervious concrete (MKGPC) and its removal capability of heavy metal ions. International Journal of Pavement Engineering 2019;0:1–12. DOI: 10.1080/10298436.2019.1667499.

[95] Bumanis G, Novais RM, Carvalheiras J, Bajare D, Labrincha JA. Metals removal from aqueous solutions by tailored porous waste-based granulated alkali-activated materials. Appl Clay Sci 2019;179:105147. DOI: 10.1016/j.clay.2019.105147.

[96] Lan T, Li P, Rehman FU, Li X, Yang W, Guo S. Efficient adsorption of Cd2+ from aqueous solution using metakaolin geopolymers. Environ Sci Pollut Res 2019;26:33555–33567. DOI: 10.1007/s11356-019-06362-w.

[97] Chen S, Qi Y, Cossa JJ, Deocleciano Salomao Dos SI. Efficient removal of radioactive iodide anions from simulated wastewater by HDTMA-geopolymer. Progress in Nuclear Energy 2019;117:103112. DOI: 10.1016/j.pnucene.2019.103112.

[98] Falah M, MacKenzie KJD, Knibbe R, Page SJ, Hanna JV. New composites of nanoparticle Cu (I) oxide and titania in a novel inorganic polymer (geopolymer) matrix for destruction of dyes and hazardous organic pollutants. Journal of Hazardous Materials 2016;318:772–782. DOI: 10.1016/j.jhazmat.2016.06.016.

[99] Yu Z, Song W, Li J, Li Q. Improved simultaneous adsorption of Cu(II) and Cr (VI) of organic modified metakaolin-based geopolymer. Arab J Chem 2020; 13:4811–4823. DOI: 10.1016/j.arabjc.2020.01.001.

[100] Hua P, Sellaoui L, Franco D, Netto MS, Luiz Dotto G, Bajahzar A, et al. Adsorption of acid green and procion red on a magnetic geopolymer based adsorbent: Experiments, characterization and theoretical treatment. Chemical Engineering Journal 2020;383:123113. DOI: 10.1016/j.cej.2019.123113.

[101] Chen H, Zhang YJ, He PY, Li CJ, Li H. Coupling of self-supporting geopolymer membrane with intercepted Cr(III) for dye wastewater treatment by hybrid photocatalysis and membrane separation. Applied Surface Science 2020;515:146024. DOI: 10.1016/j.apsusc.2020.146024.

[102] Mladenovic N, Kljajevic L, Nenadovic S, Ivanovic M, Calija B, Gulicovski J, et al. The Applications of New Inorganic Polymer for Adsorption Cadmium from Waste Water. J Inorg Organomet Polym Mater 2020;30:554–563. DOI: 10.1007/s10904-019-01215-y.

[103] da Costa Rocha AC, Scaratti G, Moura-Nickel CD, da Silva TL, Gurgel Adeodato Vieira M, Peralta RM, et al. Economical and Technological Aspects of Copper Removal from Water Using a Geopolymer and Natural Zeolite. Water Air Soil Pollut 2020;231:361. DOI: 10.1007/s11270-020-04722-8.

[104] Papa E, Mor M, Natali Murri A, Landi E, Medri V. Ice-templated geopolymer beads for dye removal. Journal of Colloid and Interface Science 2020;572:364–373. DOI: 10.1016/j.jcis.2020.03.104.

[105] Medri V, Papa E, Mor M, Vaccari A, Natali Murri A, Piotte L, et al. Mechanical strength and cationic dye adsorption ability of metakaolin-based geopolymer spheres. Applied Clay Science 2020;193:105678. DOI: 10.1016/j.clay.2020.105678.

[106] Netto MS, Rossatto DL, Jahn SL, Mallmann ES, Dotto GL, Foletto EL. Preparation of a novel magnetic geopolymer/zero–valent iron composite with remarkable adsorption performance towards aqueous Acid Red 97. Chemical Engineering Communications 2020;207:1048–1061. DOI: 10.1080/00986445.2019.1635467.

[107] Samarina T, Takaluoma E, Kazmaganbetova A. Geopolymers Supported on Inert Substrate for Phosphate Removal from Natural Waters. Proceedings of the 6th World Congress on New Technologies (NewTech'20), Virtual: AVESTIA; 2020. DOI: 10.11159/icepr20.119.

[108] Huang J, Li Z, Zhang J, Zhang Y, Ge Y, Cui X. In-situ synchronous carbonation and self-activation of biochar/geopolymer composite membrane: Enhanced catalyst for oxidative degradation of tetracycline in water. Chemical Engineering Journal 2020;397:125528. DOI: 10.1016/j.cej.2020.125528.

[109] Petlitckaia S, Barré Y, Piallat T, Grauby O, Ferry D, Poulesquen A. Functionalized geopolymer foams for cesium removal from liquid nuclear waste. Journal of Cleaner Production 2020;269:122400. DOI: 10.1016/j.jclepro.2020.122400.

[110] Lertcumfu N, Jaita P, Thammarong S, Lamkhao S, Tandorn S, Randorn C, et al. Influence of graphene

oxide additive on physical, microstructure, adsorption, and photocatalytic properties of calcined kaolinite-based geopolymer ceramic composites. Colloid Surf A-Physicochem Eng Asp 2020;602:125080. DOI: 10.1016/j.colsurfa.2020.125080.

[111] Al-Harahsheh MS, Al Zboon K, Al-Makhadmeh L, Hararah M, Mahasneh M. Fly ash based geopolymer for heavy metal removal: A case study on copper removal. Journal of Environmental Chemical Engineering 2015;3:1669–1677. DOI: 10.1016/j.jece.2015.06.005.

[112] Meher AK, Das S, Rayalu S, Bansiwal A. Enhanced arsenic removal from drinking water by iron-enriched aluminosilicate adsorbent prepared from fly ash. Desalin Water Treat 2015; 57:20944–20956. DOI: 10.1080/19443994.2015.1112311.

[113] Zhang K, Zhang D, Zhang K. Arsenic removal from water using a novel amorphous adsorbent developed from coal fly ash. Water Sci Technol 2016;73:1954–1962. DOI: 10.2166/wst.2016.028.

[114] Liu Y, Yan C, Zhang Z, Wang H, Zhou S, Zhou W. A comparative study on fly ash, geopolymer and faujasite block for Pb removal from aqueous solution. Fuel 2016;185:181–189. DOI: 10.1016/j.fuel.2016.07.116.

[115] Samarina T, Takaluoma E. New Geopolymer Adsorbents for Phosphate Removal from Diluted Solutions and their Applications. Proceedings of the 5th World Congress on New Technologies (NewTech'19), Lisbone: AVESTIA; 2019. DOI: 10.11159/ICEPR19.158.

[116] Mužek MN, Svilović S, Zelić J. Kinetic studies of cobalt ion removal from aqueous solutions using fly ash-based geopolymer and zeolite NaX as sorbents. Separation Science and Technology 2016;51:2868–2875. DOI: 10.1080/01496395.2016.1228675.

[117] Qiu J, Zhao Y, Xing J, Sun X. Fly ash-based geopolymer as a potential adsorbent for Cr(VI) removal. Desalin Water Treat 2017;70:201–209. DOI: 10.5004/dwt.2017.20493.

[118] Lee NK, Khalid HR, Lee HK. Adsorption characteristics of cesium onto mesoporous geopolymers containing nano-crystalline zeolites. Microporous Mesoporous Mat 2017;242:238–244. DOI: 10.1016/j.micromeso.2017.01.030.

[119] Maiti M, Sarkar M, Malik MA, Xu S, Li Q, Mandal S. Iron Oxide NPs Facilitated a Smart Building Composite for Heavy-Metal Removal and Dye Degradation. ACS Omega 2018;3:1081–1089. DOI: 10.1021/acsomega.7b01545.

[120] Zhang YJ, He PY, Zhang YX, Chen H. A novel electroconductive graphene/fly ash-based geopolymer composite and its photocatalytic performance. Chemical Engineering Journal 2018;334:2459–2466. DOI: 10.1016/j.cej.2017.11.171.

[121] Samiullah M, Aslam Z, Rana AG, Abbas A, Ahmad W. Alkali-Activated Boiler Fly Ash for Ni(II) Removal: Characterization and Parametric Study. Water Air Soil Pollut 2018;229:113. DOI: 10.1007/s11270-018-3758-5.

[122] Darmayanti L, Notodarmojo S, Damanhuri E, Kadja GTM, Mukti RR. Preparation of alkali-activated fly ash-based geopolymer and their application in the adsorption of copper (II) and zinc (II) ions. MATEC Web Conf 2019;276:06012. DOI: 10.1051/matecconf/201927606012.

[123] Onutai S, Kobayashi T, Thavorniti P, Jiemsirilers S. Porous fly ash-based geopolymer composite fiber as an adsorbent for removal of heavy metal ions from wastewater. Materials

Letters 2019;236:30–33. DOI: 10.1016/j.matlet.2018.10.035.

[124] Soni R, Shukla DP. Synthesis of fly ash based zeolite-reduced graphene oxide composite and its evaluation as an adsorbent for arsenic removal. Chemosphere 2019;219:504–509. DOI: 10.1016/j.chemosphere.2018.11.203.

[125] Tian Q, Sasaki K. Application of fly ash-based geopolymer for removal of cesium, strontium and arsenate from aqueous solutions: kinetic, equilibrium and mechanism analysis. Water Sci Technol 2019;79:2116–2125. DOI: 10.2166/wst.2019.209.

[126] Siyal AA, Shamsuddin MR, Rabat NE, Zulfiqar M, Man Z, Low A. Fly ash based geopolymer for the adsorption of anionic surfactant from aqueous solution. Journal of Cleaner Production 2019;229:232–243. DOI: 10.1016/j.jclepro.2019.04.384.

[127] Darmayanti L, Kadja GTM, Notodarmojo S, Damanhuri E, Mukti RR. Structural alteration within fly ash-based geopolymers governing the adsorption of Cu2+ from aqueous environment: Effect of alkali activation. J Hazard Mater 2019;377:305–314. DOI: 10.1016/j.jhazmat.2019.05.086.

[128] Shao N, Tang S, Li S, Chen H, Zhang Z. Defective analcime/geopolymer composite membrane derived from fly ash for ultrafast and highly efficient filtration of organic pollutants. Journal of Hazardous Materials 2020;388:121736. DOI: 10.1016/j.jhazmat.2019.121736.

[129] Abdelrahman EA, Abou El-Reash YG, Youssef HM, Kotp YH, Hegazey RM. Utilization of rice husk and waste aluminum cans for the synthesis of some nanosized zeolite, zeolite/zeolite, and geopolymer/zeolite products for the efficient removal of Co (II), Cu(II), and Zn(II) ions from aqueous media. Journal of Hazardous Materials 2021;401:123813. DOI: 10.1016/j.jhazmat.2020.123813.

[130] Mondal SK, Welz A, Rezaei F, Kumar A, Okoronkwo MU. Structure-Property Relationship of Geopolymers for Aqueous Pb Removal. ACS Omega 2020;5:21689–21699. DOI: 10.1021/acsomega.0c02591.

[131] Novais RM, Carvalheiras J, Seabra MP, Pullar RC, Labrincha JA. Highly efficient lead extraction from aqueous solutions using inorganic polymer foams derived from biomass fly ash and metakaolin. J Environ Manage 2020;272:111049. DOI: 10.1016/j.jenvman.2020.111049.

[132] Zhang YJ, He PY, Yang MY, Kang L. A new graphene bottom ash geopolymeric composite for photocatalytic H2 production and degradation of dyeing wastewater. International Journal of Hydrogen Energy 2017;42:20589–20598. DOI: 10.1016/j.ijhydene.2017.06.156.

[133] Zhang YJ, Yang MY, Zhang L, Zhang K, Kang L. A new graphene/geopolymer nanocomposite for degradation of dye wastewater. Integrated Ferroelectrics 2016;171:38–45. DOI: 10.1080/10584587.2016.1171178.

[134] Runtti H, Luukkonen T, Niskanen M, Tuomikoski S, Kangas T, Tynjälä P, et al. Sulphate removal over barium-modified blast-furnace-slag geopolymer. Journal of Hazardous Materials 2016;317:373–384. DOI: 10.1016/j.jhazmat.2016.06.001.

[135] He PY, Zhang YJ, Chen H, Liu LC. Development of an eco-efficient CaMoO4/electroconductive geopolymer composite for recycling silicomanganese slag and degradation of dye wastewater. Journal of Cleaner Production 2019;208:1476–1487. DOI: 10.1016/j.jclepro.2018.10.176.

[136] Chen F, Wang K, Shao L, Muhammad Y, Wei Y, Gao F, et al.

Synthesis of Fe2O3-modified porous geopolymer microspheres for highly selective adsorption and solidification of F− from waste-water. Composites Part B: Engineering 2019;178:107497. DOI: 10.1016/j.compositesb.2019.107497.

[137] Liu Z, Li L, Shao N, Hu T, Han L, Wang D. Geopolymerization enhanced hydrothermal synthesis of analcime from steel slag and CFBC fly ash and heavy metal adsorption on analcime. Environ Technol 2020;41:1753–1765. DOI: 10.1080/09593330.2018.1545805.

[138] Wang K, Lei H, Muhammad Y, Chen F, Gao F, Wei Y, et al. Controlled preparation of cerium oxide loaded slag-based geopolymer microspheres (CeO2@SGMs) for the adsorptive removal and solidification of F− from acidic waste-water. Journal of Hazardous Materials 2020;400:123199. DOI: 10.1016/j.jhazmat.2020.123199.

[139] Yan S, He P, Jia D, Wang Q, Liu J, Yang J, et al. Synthesis of novel low-cost porous gangue microsphere/geopolymer composites and their adsorption properties for dyes. International Journal of Applied Ceramic Technology 2018;15:1602–14. DOI: https://doi.org/10.1111/ijac.13045.

[140] Al-Zboon KK, Al-smadi BM, Al-Khawaldh S. Natural Volcanic Tuff-Based Geopolymer for Zn Removal: Adsorption Isotherm, Kinetic, and Thermodynamic Study. Water Air Soil Pollut 2016;227:248. DOI: 10.1007/s11270-016-2937-5.

[141] Minelli M, Medri V, Papa E, Miccio F, Landi E, Doghieri F. Geopolymers as solid adsorbent for CO2 capture. Chemical Engineering Science 2016;148:267–274. DOI: 10.1016/j.ces.2016.04.013.

[142] Li C, He Y, Tang Q, Wang K, Cui X. Study of the preparation of CdS on the surface of geopolymer spheres and photocatalyst performance. Materials Chemistry and Physics 2016;178:204–210. DOI: 10.1016/j.matchemphys.2016.05.013.

[143] Abdelrahman EA, Subaihi A. Application of Geopolymers Modified with Chitosan as Novel Composites for Efficient Removal of Hg(II), Cd(II), and Pb(II) Ions from Aqueous Media. J Inorg Organomet Polym Mater 2020;30:2440–2463. DOI: 10.1007/s10904-019-01380-0.

[144] Demir F, Moroydor Derun E. Usage of gold mine tailings based geopolymer on Cu 2+ adsorption from water. Main Group Chemistry 2019;18:467–476. DOI: 10.3233/MGC-190801.

[145] Wang J, Zhao Y, Zhang P, Yang L, Xu H, Xi G. Adsorption characteristics of a novel ceramsite for heavy metal removal from stormwater runoff. Chinese Journal of Chemical Engineering 2018;26:96–103. DOI: 10.1016/j.cjche.2017.04.011.

[146] Paranaque JEL, Maguyon-Detras MC, Migo VP, Alfafara CG. Chromium removal from chrome-tannery effluent after alkaline precipitation by adsorption using municipal solid waste-derived activated biochar. 26th Regional Symposium on Chemical Engineering (rsce 2019), vol. 778, Bristol: Iop Publishing Ltd; 2020, p. 012134. DOI: 10.1088/1757-899X/778/1/012134.

[147] Panda L, Jena SK, Rath SS, Misra PK. Heavy metal removal from water by adsorption using a low-cost geopolymer. Environ Sci Pollut Res 2020;27:24284–24298. DOI: 10.1007/s11356-020-08482-0.

[148] Bakharev T. Thermal behaviour of geopolymers prepared using class F fly ash and elevated temperature curing. Cement and Concrete Research 2006;36:1134–1147. DOI: 10.1016/j.cemconres.2006.03.022.

[149] Luukkonen T, Tolonen E-T, Runtti H, Kemppainen K, Perämäki P,

Rämö J, et al. Optimization of the metakaolin geopolymer preparation for maximized ammonium adsorption capacity. J Mater Sci 2017;52:9363–9376. DOI: 10.1007/s10853-017-1156-9.

[150] Siyal AA, Shamsuddin MR, Khan MI, Rabat NE, Zulfiqar M, Man Z, et al. A review on geopolymers as emerging materials for the adsorption of heavy metals and dyes. Journal of Environmental Management 2018;224: 327–339. DOI: 10.1016/j.jenvman.2018.07.046.

[151] Luukkonen T, Heponiemi A, Runtti H, Pesonen J, Yliniemi J, Lassi U. Application of alkali-activated materials for water and wastewater treatment: a review. Rev Environ Sci Biotechnol 2019;18:271–297. DOI: 10.1007/s11157-019-09494-0.

[152] Luukkonen T, Věžníková K, Tolonen E-T, Runtti H, Yliniemi J, Hu T, et al. Removal of ammonium from municipal wastewater with powdered and granulated metakaolin geopolymer. Environmental Technology 2018;39: 414–423. DOI: 10.1080/09593330.2017.1301572.

[153] Tan TH, Mo KH, Ling T-C, Lai SH. Current development of geopolymer as alternative adsorbent for heavy metal removal. Environmental Technology & Innovation 2020;18:100684. DOI: 10.1016/j.eti.2020.100684.

[154] Ducman V, Korat L. Characterization of geopolymer fly-ash based foams obtained with the addition of Al powder or H2O2 as foaming agents. Materials Characterization 2016; 113:207–213. DOI: 10.1016/j.matchar.2016.01.019.

[155] Landi E, Medri V, Papa E, Dedecek J, Klein P, Benito P, et al. Alkali-bonded ceramics with hierarchical tailored porosity. Applied Clay Science 2013;73:56–64. DOI: 10.1016/j.clay.2012.09.027.

[156] Li Q, He Y, Xu M, Liu J, He M, Huang L, et al. Study on the removal of Ca2+ and Mg2+ in water by the geopolymer-based inorganic membrane 2017;48:01215–20. DOI: 10.3969/j.issn.1001-9731.2017.01.039.

[157] Ge Y, Yuan Y, Wang K, He Y, Cui X. Preparation of geopolymer-based inorganic membrane for removing Ni2+ from wastewater. Journal of Hazardous Materials 2015;299:711–718. DOI: 10.1016/j.jhazmat.2015.08.006.

[158] Davidovits J. 30 years of successes and failures in geopolymer applications. Market trends and potential breakthroughs. Geopolymer 2002 Conference 2002:1–16.

[159] Cheng TW, Lee ML, Ko MS, Ueng TH, Yang SF. The heavy metal adsorption characteristics on metakaolin-based geopolymer. Applied Clay Science 2012;56:90–96. DOI: 10.1016/j.clay.2011.11.027.

[160] El-Eswed B, Alshaaer M, Yousef RI, Hamadneh I, Khalili F. Adsorption of Cu(II), Ni(II), Zn(II), Cd (II) and Pb(II) onto Kaolin/Zeolite Based- Geopolymers. Advances in Materials Physics and Chemistry 2013;2: 119–125. DOI: 10.4236/ampc.2012.24B032.

[161] Luukkonen T, Sarkkinen M, Kemppainen K, Rämö J, Lassi U. Metakaolin geopolymer characterization and application for ammonium removal from model solutions and landfill leachate. Applied Clay Science 2016;119:266–276. DOI: 10.1016/j.clay.2015.10.027.

[162] Bai C, Colombo P. Processing, properties and applications of highly porous geopolymers: A review. Ceramics International 2018;44:16103–16118. DOI: 10.1016/j.ceramint.2018.05.219.

[163] Bai C, Franchin G, Elsayed H, Zaggia A, Conte L, Li H, et al. High-

porosity geopolymer foams with tailored porosity for thermal insulation and wastewater treatment. Journal of Materials Research 2017;32:3251–3259. DOI: 10.1557/jmr.2017.127.

[164] Huang X, Zhao H, Hu X, Liu F, Wang L, Zhao X, et al. Optimization of preparation technology for modified coal fly ash and its adsorption properties for Cd2+. J Hazard Mater 2020;392: 122461. DOI: 10.1016/j.jhazmat.2020.122461.

[165] Zhang YJ, Liu LC, Xu Y, Wang YC, Xu DL. A new alkali-activated steel slag-based cementitious material for photocatalytic degradation of organic pollutant from waste water. Journal of Hazardous Materials 2012;209–210: 146–150. DOI: 10.1016/j.jhazmat.2012.01.001.

[166] Samarina T, Takaluoma E. Ammonia Recovery from Wastewaters by Adsorption with subsequent Transmembrane Chemical Absorption. Proceedings of the 6th World Congress on New Technologies (NewTech'20), Virtual: AVESTIA; 2020. DOI: 10.11159/icepr20.146.

[167] Takaluoma E, Samarina T, Mwandila G, Kabondo Leonard L, Nyirenda Kawunga K, Mwaanga P. Selective Recovery of Copper and Cobalt from Mine Effluent. IMWA 2020 Mine water solutions, New Zealand: IMWA; 2020, p. 181186.

[168] Lata S, Singh PK, Samadder SR. Regeneration of adsorbents and recovery of heavy metals: a review. Int J Environ Sci Technol 2015;12:1461–1478. DOI: 10.1007/s13762-014-0714-9.

Chapter 2

The Physical and Mechanical Characteristics of Geopolymers Using Mine Tailings as Precursors

Petrica Vizureanu, Dumitru Doru Burduhos Nergis, Andrei Victor Sandu, Diana Petronela Burduhos Nergis and Madalina Simona Baltatu

Abstract

Mine tailings are waste materials that resulted from the extraction and processing of raw materials to form mineral products. These dusty particles present negative environmental effects after being deposited in different types of dumping areas. Based on the circular economy concepts and the presently pushing need of identifying wastes as a potential replacement for natural resources, this chapter aims to present the physical (density, microstructure) and mechanical (compressive strength, flexural strength) characteristics of different types of geopolymers which use mine tailings as precursors or blended systems (mixes of different raw materials). The main reasons of approaching this topic are the need to decrease the consumption of natural resources, reduce environmental pollution and create an economic system aimed to capitalize the mining wastes. Accordingly, this chapter includes information regarding the availability of this waste and its potential utilization as a raw material in civil engineering applications. Therefore, reports of specific agencies and multiple research studies which approach tailing based geopolymers or blended systems have been summarized.

Keywords: geopolymers, mine tailings, ecofriendly materials, fusion activation, leaching, reinforced structures, flexural strength, compressive strength

1. Introduction

Nowadays, globalization generates large amounts of waste that significantly affects the storage areas and the surrounding environment. At the same time, the civil engineering sector is experiencing an exponential development process, which increases the demand for building materials and usable space. Therefore, the need to obtain new materials with lower exploitation costs and natural resources consumption became primary. One solution that has been intensively studied in the last past year, especially in this sector, consists of the development of environmentally friendly materials through a mechanism called geopolymerisation (**Figure 1**). The resulting materials, the geopolymers, consist of a tetrahedra network of aluminates (AlO_4) and silicates (SiO_4), chemically balanced by alkali ions of K^+, Na^+ or Li^+ [1]. The geosynthesis manifests itself in nature in great abundance. The Earth consists

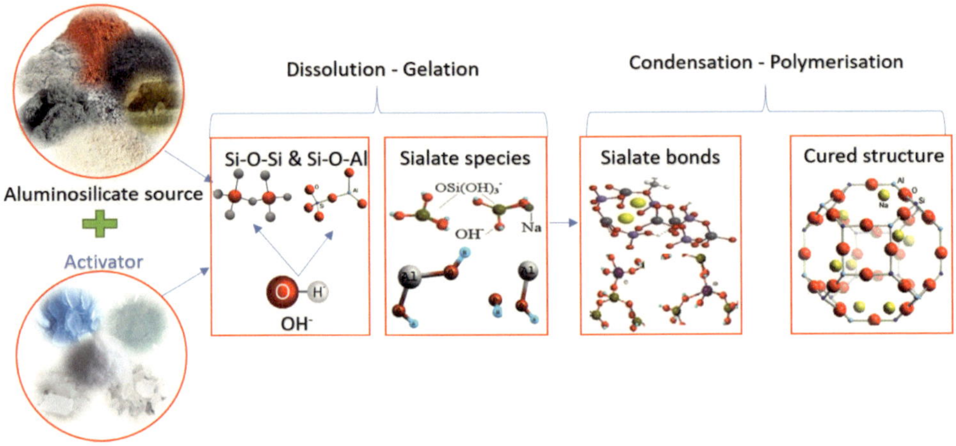

Figure 1.
Geopolymerisation process.

of 55% of volume, from siloxo-sialates and sialates, but only 12% pure silica or quartz. The geosynthesis process is based on changes induced in the crystallography of the silica backbone by the aluminum ion (6-fold or 4-fold coordination) and on the chemical changes produced by the same aluminum ion. This geosynthesis process is based on changes in crystallography of the silica backbone by the aluminum ion (VI-fold or IV-fold coordination) and the changes on the chemical part made by the same aluminum ion [2]. In this study, geopolymers refer to alkali activated materials obtained through the geopolymerisation reaction.

Currently, these materials are used in multiple industries, starting from civil engineering applications, up to medicine and spaces industry [3–5]. Therefore, there is high interest in the development of new geopolymers that possess higher properties than conventional materials (such as concretes with ordinary Portland cement or ceramics that use natural resources for their synthesis) [6]. However, the main zone of geopolymeric technology application is in the development of low CO_2 construction materials, mainly as an alternative to Portland-based (calcium silicate) cement [7, 8]. Being a performant material is not enough for the market, where it cannot go without a real demand for materials with such characteristics.

Nevertheless, the geopolymers preference over conventional materials is also supported by the soil decontamination potential of these materials, which is mainly related to the possibility to use waste as precursors or as reinforcing elements (**Figure 2**) [3, 9, 10]. Considering these characteristics, i.e. high properties and positive impact on the environment, it can be stated that this cycle is energy saving, natural resources conserver and waste-reducing [11, 12]. Continuing into this idea, using waste it's economically friendly and, first of all, cheaper. Therefore, up to now, waste such as coal ash, red mud, slags, rice husk ash etc. have been investigated as geopolymers precursors [13]. However, the possibility of using waste for geopolymers manufacturing is mainly constrained by the availability of the aluminosilicate source [14, 15]. Based on this limitation, a considerable source for geopolymerisation has been identified, i.e. mine tailings. Moreover, the use of this waste was also encouraged by the fact that mine tailings can be used in blended geopolymers (two or more types of aluminum and silicone rich powders are used for the geopolymers' synthesis). Therefore, when the properties of the final product aren't suitable for the specific application, i.e. when tailored properties are required, usually, the structure of the geopolymers must be reinforced with different types of particles which will contribute to their mechanical characteristics [16–18].

Figure 2.
The main components of waste-based geopolymers.

During the last decades, numerous studies have been focused on using mine tailings as precursors or blended component in geopolymers. Therefore, a compressive presentation of the obtained results, i.e. state-of-the-art review, could be a strong encouraging point for further research on this topic and scientific progress. Accordingly, this chapter includes information regarding the availability of this waste and its potential utilization as a raw material in civil engineering applications. Therefore, reports of specific agencies have been summarized and multiple research studies which approach tailing based geopolymers or blended systems are reviewed, and future research objectives have been presented accordingly.

2. Resources and production methods

The resources of an element consisting of the available amount in the earth's crust, the oceans and the atmosphere, which could be identified as extractible. The part of these resources, which can be exploited in economic conditions at a given time, represents the reserves. The boundary between resources and reserves varies over time depending on the economic and technological factors of exploitation and the strategy of different states and industrial groups. The average concentrations of the main elements from the earth crust are presented in **Table 1**. As can be observed,

Chemical element	O	Si	Al	Fe	Ca*	Na	K	Mg**	Oth. Elem.***
%, wt.	46.6	27.7	8.1	5.0	3.6	2.8	2.6	2.1	1.5

*Ca – calcium.
**Mg – magnesium.
***Oth. Elem.- other chemical elements.

Table 1.
The main chemical elements from the earth crust [19].

eight elements represent almost 98.5% of the earth's crust. The base elements for geopolymerisation, i.e. oxygen (O), aluminum (Al) and silicon (Si), are situated in the first three positions and occupies almost 82% by weight from the entire earth's crust. However, because one more element (sodium (Na) and/or potassium (K) or lithium (Li)) is needed for activation, the total percentage is increased up to 88%. Moreover, considering the fact that aluminum, iron (Fe) and copper are the most extracted elements, large quantities of those goes to the mine tailing dumps, becoming a resource for geopolymers.

In order to obtain different products, the extracted resources go through a cycle of physical, chemical and mechanical processing called the circuit of materials in the production process. After extraction, the raw material undergoes a series of physical and chemical transformations until the final product is obtained (aluminum, iron, copper ingots etc.). During this transformation process, different parts from the raw material are lost, therefore, those volumes of extracted materials go to waste. Extraction, processing and manufacturing of materials require large amounts of energy and therefore production costs are highly dependent on the price of energy. From an energy point of view, organic materials are much more cost-effective, because their synthesis and shaping require much lower energy consumption than metals or ceramic materials. Therefore, to reduce the production costs, the waste can be recycled and capitalized by converting it into raw materials for other products.

In the development of one technology, one material is often substituted for another, for economic or performance reasons. Thus, the car body was originally made of wood - a light material, existing in nature. Then the wood was replaced with steel sheet, a heavier but more resistant material, with controllable properties and easy to process into complex shapes. To reduce energy consumption, we switched to lighter materials. This has led to the use of thin sheets of high-strength steel, as well as very light unidirectional composite materials made of carbon fibers embedded in organic polymers.

2.1 Recycled raw materials

Anything that is not recycled or recovered from waste represents a loss of raw materials and other production factors used in the chain, in terms of production, transport and product consumption, respectively. Therefore, the environmental impacts of these secondary products are significantly higher than those associated exclusively with the effects produced during the deposition in waste dumps.

Directly or indirectly, waste affects our health and well-being in many ways: methane gas contributes to climate change, air pollutants are released into the atmosphere, drinking water sources are contaminated, crops grow on contaminated land, and fish ingest toxic chemicals, after which they reach our plates.

2.2 Coal ash

The main waste of interest to the industry, resulting from the burning of coal in thermal power plants, is the ash from thermal power plants. The interest shown by researchers in a multitude of fields is mainly due to its hydraulic properties and chemical composition (high content of oxides of silicon, aluminum, calcium and iron). Thus, it has been shown that these pozzolanic powders can become raw materials for the manufacture of technically and economically competitive materials.

Coal-ash (fly ash and bottom ash) is a result of coal burning in thermal power plants for producing energy, which is an important polluter of the environment and lands in the close areas of power plants and coal ash store units/facilities. Since the need for energy is increasing, the power plants will produce more coal-ash which is estimated approximately to 776MT per annum [20]. Because of this enormous quantity, it is necessary to make alternative technology to reduce environmental impact.

Coal ash is a silico-aluminum or low calcium material that can be used in many applications, primarily in the construction industry, e.g. concrete, pavements, recipients for containment and immobilization of radioactive wastes, refractory ceramics and because of its elemental structure, a source of geopolymerisation reaction. Due to these properties fly ash gives great mechanical strength and a good fire and chemical resistance, which influenced the researchers to find a different application of fly ash. Coal ash, especially, fly ash has been intensively studied in the geopolymers technology, therefore, this waste already presents a high interest for the researcher, and it will not be evaluated/presented intensively here [21].

2.3 Mine tailings

The first stage of ore processing consists of hard rock blocks (ore) crushing and grinding up to particles with a diameter of a few centimeters or even micrometers. Secondly, the use phase is separated from the gangue part by specific means (depending on the type of ores and the used extraction technology). Mineral separation is achieved by various methods, namely: gravimetric; magnetic; electric. The surface properties of the mineral phases can also be used in the separation process. Accordingly, the products that resulted from ore processing are the concentrate and the tailings. The concentrate is processed further until the desired metal is obtained, while the tailings are deposited in different types of dumps/facilities.

A tailings dump can be defined as "the site of surface storage and the deposit of tailings extracted from the mine or tailings resulting from mechanical preparation operations". Accordingly, tailings ponds are excavated land surfaces in which liquid waste with a high content of suspensions is deposited, in order to sediment them, while mining tailings dumps are surfaces on which the material resulting from the excavation of non-metalliferous and metalliferous ores has been deposited.

As a result of the way the excavated and piled material is deposited, the piles are in the form of mounds with the appearance of a pyramid trunk, which have an upper part, more or less horizontal, which constitutes the plateau part and is bordered around the slopes.

The mineralogical and granulometric composition of the dumped material is strongly influenced by the dependence on the geological and lithological structure of the territory studied.

Under conditions of Neogene sedimentary formations composed of intercalations of fine sandy marls, sands, gravels, clays interspersed with bundles of layers of variable thickness, the uncovered and non-selectively deposited piled material leads to a mineralogical and granulometric structure highly variable from one pile to another and especially inside the same dump.

Mining has a strong influence on the environment, through various forms and ways affecting most environmental factors, namely: air, water, flora, fauna, landscape and human settlements, cultural heritage, population health, agriculture. Among the most important influences are [22]:

- affecting large areas of land that can no longer be used for other purposes for a very long time;
- pollution of soil, groundwater and surface water with various compounds solubilized by the action of rainwater;
- in the case of exploitation of materials containing sulfides, the phenomenon of acid drainage is amplified;
- visual impact (unattractive landscapes, dust picked up by the wind affects the traffic visibility etc.).

Waste recycling brings many benefits (**Figure 3**), the most important being: conservation of natural resources, reduction of storage space, protection of the environment and recovery of materials deposited in dumps by developing new materials.

The physical and the chemical characteristics of the tailings vary considerably, depending on the mineralogical and geochemical composition, sedimentation characteristics, specific gravity of the particles, rheology and viscosity, hydraulic permeability and conductivity, the evolution of the cementation process, the chemical composition of the water in the pores, the degree of contamination of the external

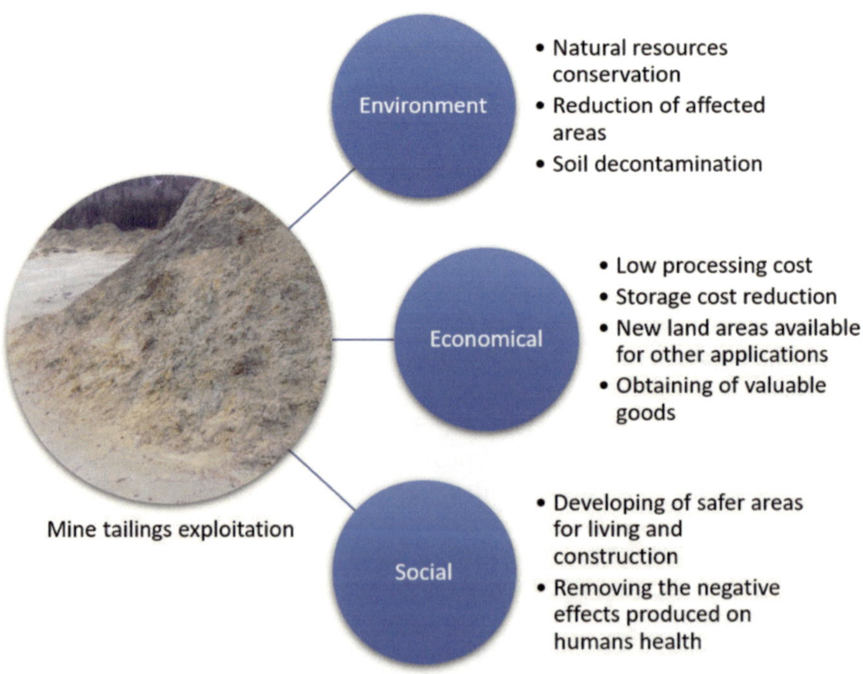

Figure 3.
Advantages of mine tailing recycling.

surface or underground environment etc. The mineralogical and geochemical composition of the solid material is strictly dependent on the paragenetic characteristics of the processed deposits. This aspect gives the tailings ponds uniqueness because there aren't two ore deposits with the same characteristics. On top of, depending on the quality and performance of the technology used in the separation process, tailings ponds can sometimes contain large amounts of metal sources.

The effects produced by the mine tailings facilities on the soil are, also, related to the raising type of storage (**Figure 4**). As can be seen, over 40% of the tailing dumps worldwide are upstream, mostly, because this is the cheapest design, unfortunately, this is also the most susceptible to failure design, which can result in huge environmental consequences [23]. Downstream is the following inline, being used at over 30% of the tailing's facilities. This design eliminates the disadvantages specific to the upstream designs, yet, the production cost is higher and the occupied space is considerably increased. Moreover, the amount of concrete building materials is significantly higher. Centerline design is a mix between upstream and downstream, this shows a lower failure coefficient than upstream and can be realized with fewer materials than downstream. The designs used in lower percentage (single-stage, dry stack, other) are usually specific to small mines, in terms of the extracted volume, while the in-pit method requires an empty/closed mine that can be filled with the resulted waste. A simplified schematic representation of the upstream, downstream and centerline design can be seen in **Figure 5**.

Although new construction designs have been developed, due to the fact that mines producing greater amounts of mine tailings, as lower grades of ore must be processed, the waste facilities are filled overcapacity, therefore, serious tailings dam failure occur (until 2027, globally, more than 15 catastrophic failures are predicted) [24]. Accordingly, by introducing and encouraging the use of mine tailings in geopolymerisation technology, a convenable source of raw materials for civil construction applications will be developed, while the requirements for a circular economy of the mining sector will be fulfilled.

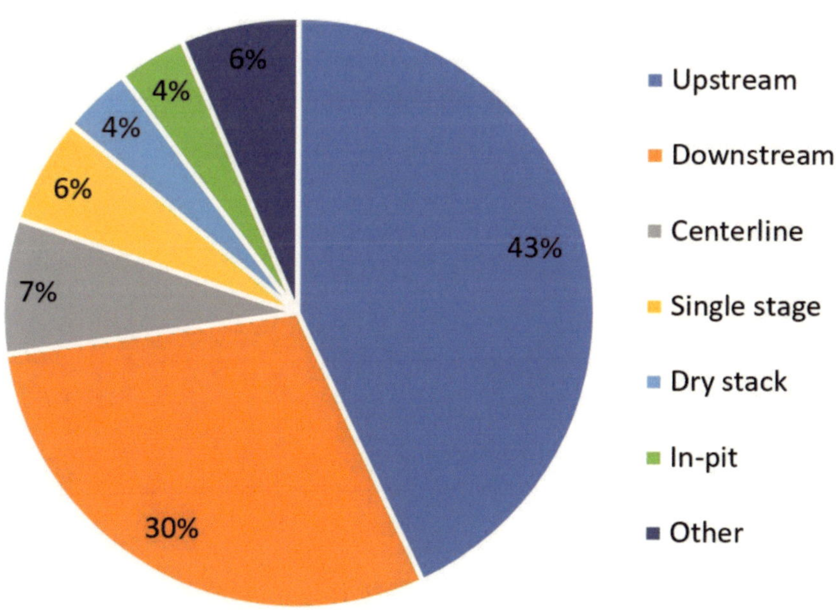

Figure 4.
Tailings storage facilities distribution, depending on the construction design type [23].

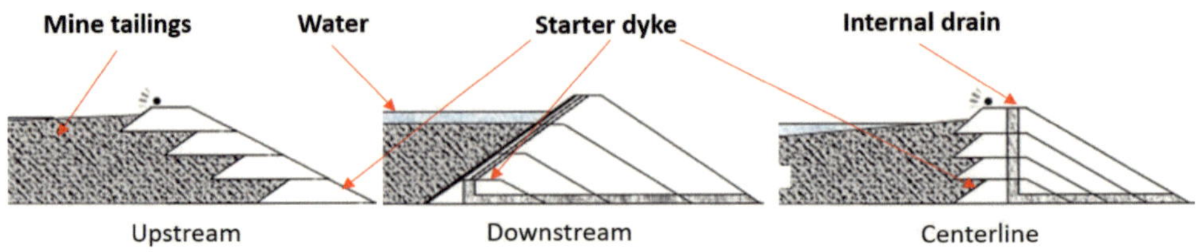

Figure 5.
Schematic representation of the construction design of tailings facilities [23].

Only in Romania, there are over 1100 tailings dumps and industrial landfills, distributed on the territory of 29 counties, respectively 13 counties for waste dumps (**Figure 6**). Of these, over 990 come from mining activities, while about 190 are located near protected areas. Moreover, more than 40 of them present serious stability problems [25]. Moreover, according to the report published by the Ministry of Economy, Romania is the country with the highest percentage (over 85%, the average in Europe being 25%) of waste resulting from the extractive industry.

Currently, the global stored volume of mine tailings is close to 55 billion cubic meters, and an increase of 23% is expected until 2025 [26]. Accordingly, the use of mine tailings in concrete can support the conservation of natural resources, specific to these activities, for 4 to 5 years [27].

2.3.1 Cooper mine tailings

Considering the fact that copper tailings are available in large volumes worldwide, and those increase considerable every day [25, 28], multiple authors focused their study on introducing these types of aluminosilicates in geopolymers technology, as precursors or partial replacements of conventional resources. Furthermore, this section aims to summarize the results obtained in this study and the main parameters that influence the feasibility of mine tailings synthetization in geopolymers.

Paiva et al. [29] developed geopolymers based on two types of metakaolin which incorporates fine particles of high-sulfidic Mine Tailings (MT) which comes from a copper and zinc mine. According to their study, the compressive strength of the geopolymers decreases with the increase of the substitution percentage despite the

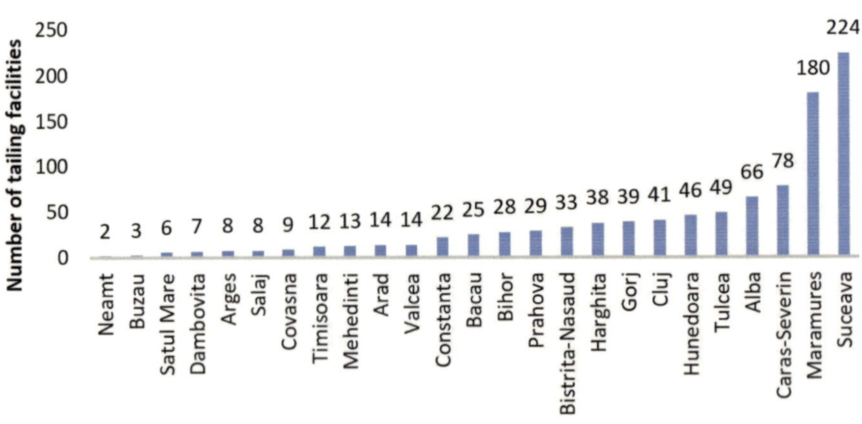

Figure 6.
Distribution of mine tailing facilities on Romania territory [25].

curing temperature. Therefore, after replacing 38% of MetaKaolin (MK) mass with mine tailings a decrease of 44% of compressive strength was obtained, while after 50% of metakaolin was replaced a decrease of 69% was observed, for the samples cured at room temperature (**Figure 7**). Based on the XRD evaluation (phases identified: Pyrite FeS_2, Anhydrite $Ca(SO_4)$, Caldecahydrite $CaAl_2O_4\ 10H_2O$, Quartz SiO_2), they stated that the analyzed mine waste cannot be used as a precursor for geopolymerisation because no reactive aluminosilicates are present in its composition. However, it can be used in blended systems, even its incorporation results in a more compact structure (bulk density increases from 1.7 to 1.9) with lower mechanical characteristics. Moreover, when Blast Furnace Slag (BFS) was used as a precursor, the bulk density decreases and so does the compressive strength, however, better values can be obtained by curing at high temperature (**Figure 8**).

The structural analysis of a cooper-barium mine tailing activated with 10 M NaOH for geopolymers synthesis exhibits a heterogeneous matrix with a partially dissolved structure, full of voids and unreacted particles (**Figure 9**).

The reactivity of mine tailings in alkaline activators was reported by Obenaus-Emler et al. [30]. According to their publication, the amount of dissolved species from metakaolin reaches 80%, for granulated furnace slag the value was close to 60%, while for copper mine tailing the dissolved amount was lower than 5%. However, better results were obtained after increasing the curing time or the NaOH concentration (**Table 2**). The compressive strength of the obtained geopolymers seems to depend on the same parameters, accordingly, an increase of 40% was obtained for samples cured at 60°C (compared with those cured at room temperature), while by increasing the concentration of water glass from 10–30%, four times higher compressive strength was obtained for room temperature cured samples, and 6 times higher for 60°C cured samples, respectively. Moreover, one more parameter that showed promising results on improving mechanical characteristics was the addition of finer particles or BFS, however, this also affects the pore size distribution of the hardened product.

In another study, Ahmari et al. [31] successfully synthesized geopolymers with copper mine tailings as precursors. According to their study, satisfying compressive strength can be obtained by customizing the curing parameters and the activator (**Figure 10**). The samples synthesized with a 15 M NaOH solution exhibit the optimum value when cured at 90 °C, while for 5 M NaOH and 10 M NaOH, 75°C was the optimum temperature.

Moreover, by introducing sodium silicate in the composition, the highest compressive strength can be obtained at SiO_2 to Na_2O equal to 1, for the geopolymer activated with 10 M NaOH and cured for 7 days at 60°C (sample 1SS). When the ratio is increased the mechanical characteristics will decrease because this solution

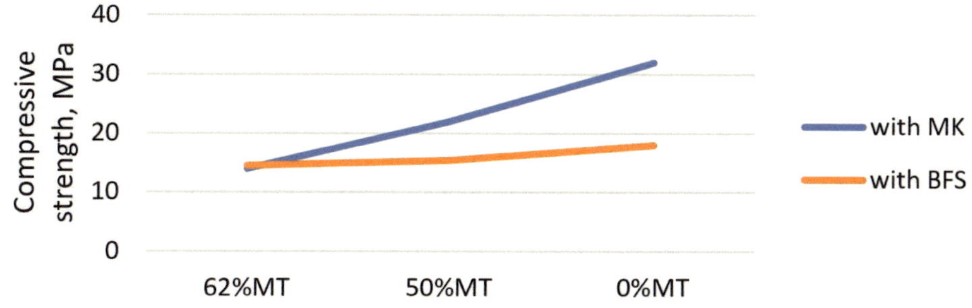

Figure 7.
The effect of precursors replacing with copper mine tailings on compressive strength (- approximate value).*

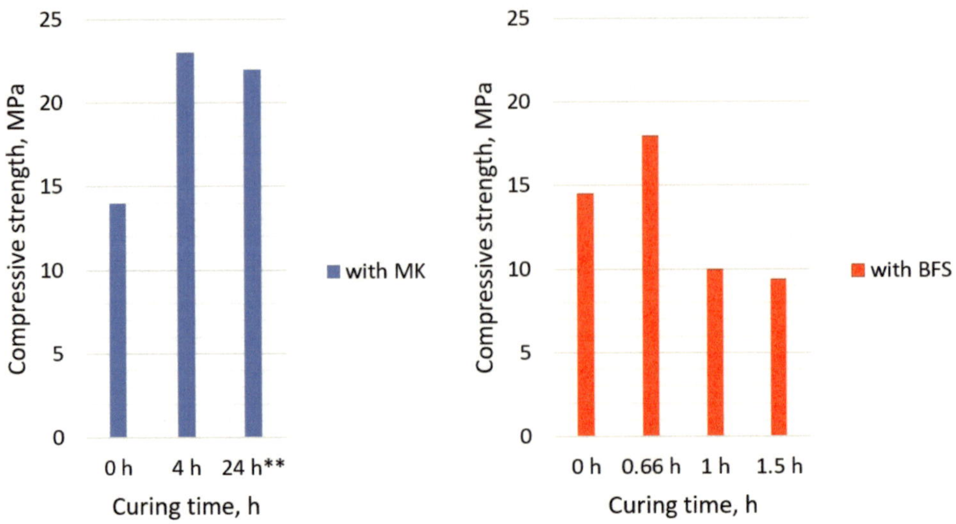

Figure 8.
*Compressive strength of 50°C cured samples for different periods (** - for sealed samples).*

Figure 9.
The microstructure of cooper-barium mine tailing based geopolymer.

provides the amount of silicon needed in the early stages of geopolymerisation, but in too high concentrations it prevents water evaporation, while encapsulates the particles of the raw material in a film that does not allow the catalyst to advance. Also, for 10 M NaOH activate geopolymers, cured at 90°C, the addition of sodium aluminate in a mass ratio of 1.25 (sodium aluminate/NaOH solution) showed an impressive increase of compressive strength from ≈6 MPa to ≈17 MPa (sample 1SA). The activator and curing parameters also affect the microstructure of the obtained samples, consequently, a denser matrix was observed at higher NaOH concentration

Raw material	Absolute amount dissolved, mg/l					
	Ca	Mg	Al	Si	Fe	S
MT, 5 M NaOH	4.4	0.3	2.2	16	3.7	97.2
MT, 10 M NaOH	13	9.2	3.1	28	37	70
MK, 10 M NaOH	—	—	2122	3148	—	—
BFS, 10 M NaOH	21.8	2.3	603	2450	1.9	880

Table 2.
The amount of dissolved species after 24 h leaching.

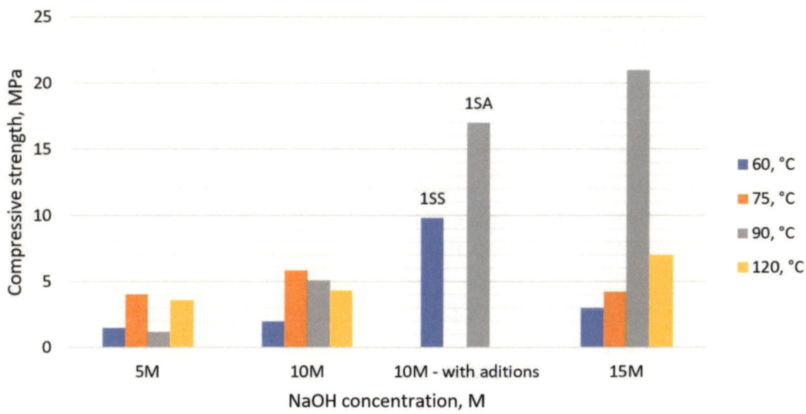

Figure 10.
Activator and curing parameters effects on compressive strength in cooper mine waste-based geopolymers.

or curing temperature. As can be seen from the graph, better results can be obtained by curing the 1SS samples at 90°C, or by increasing the molar concentration of the NaOH solution from 10 M to 15 M, in the case of geopolymers with additions.

According to another study [32], if appropriate manufacturing conditions (NaOH concentration, initial water content, forming pressure, and curing parameters) are selected, the copper mine tailings are feasible to produce eco-friendly bricks by a simple three steps method: (i) mixing the mine tailing with the activator, (ii) compress the mixture in a mold at a specific pressure, (iii) cure the products at slightly elevated temperature.

2.3.2 Tungsten mine tailings

Usually, tailings from many tungsten extraction facilities contain only small quantities of heavy metals and fatty acids, therefore, those aren't classified as class A facilities. However, due to their large amount available (the production of tungsten reached 87,000 metric tons in 2015, therefore, the amount of waste is even higher) long-term processing and managements solution must be developed [33].

Tungsten Waste-based Geopolymers (TWG) has been successfully synthesized by mixing the cementitious powder with different percentages of calcium hydroxide and mixes of NaOH and waterglass solution. According to previous studies, this type of geopolymers has good adhesion properties, low water absorption and high mechanical characteristics. In their study, Pacheco-Torgal et al. [34], evaluated the influence of aggregates (sand) and calcium hydroxide ($Ca(OH)_2$) introduction in the matrix of TWG on compressive strength evolution. The optimum $Ca(OH)_2$

concentration was 10%, while the ratio of 1 for sand to binder mass showed the most promising results. Moreover, when water to sodium hydroxide ratio was evaluated (**Figure 11a**) it was observed that higher NaOH concentrations result in better compressive strength, while higher water content badly affects the mechanical strength evolution (**Figure 11b**).

In another study [35], the authors evaluated the adhesion characteristics of tungsten-based geopolymers to the surface of pretreated OPC concrete. According to their study, the samples exhibit a high bond strength and monolithic failure even after 1 day of curing. Moreover, they stated that this type of geopolymers are three times stronger than C30/37 strength class OPC concrete, in terms of abrasion resistance, and, almost the same difference, in terms of sulfuric acid resistance. This can be related to the low unrestrained shrinkage and capillary water absorption coefficients, respectively.

2.3.3 Gold mine tailings

Considering the fact that the majority of the processed gold comes from the open querries, and their activities include substrate removal and grounding of high amounts of rock (almost 20 tones of waste is generated for a standard 18-karat ring). Gold products obtained following mining activities are through the most tailing generative.

Waste resulted from gold mining activities was also approached in geopolymers obtaining. Kiventera et al. [36] synthesized Gold Mine Tailing (GMT) based geopolymers or blended systems of GMT and BFS activated with different NaOH solutions. According to their study, due to the fact that the water quantity constrains the efficiency of the polycondensation stage, the workability was affected by the molarity of NaOH solution and solid to liquid ratio. Additionally, it was observed that a higher water quantity can result in multiples cracks and pores which

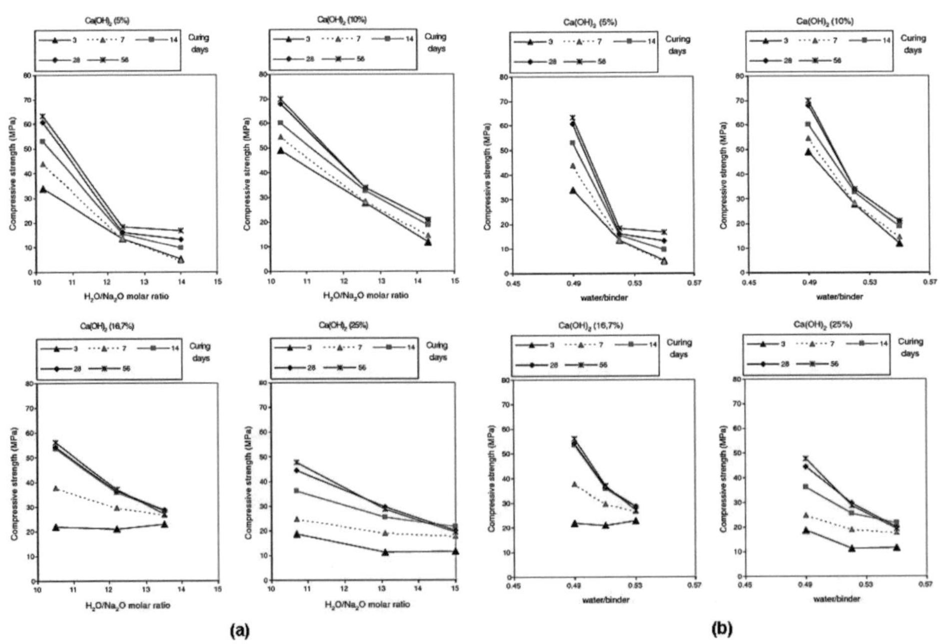

Figure 11.
The influence of different parameters on compressive strength of TWG, depending on curing time: a) water to sodium hydroxide mass ratio; b) solid to liquid ratio; [34].

significantly affects the mechanical characteristics of the samples. Therefore, for optimal properties, a high NaOH quantity must be dissolute in the binder because the Na/Al and Na/Si ratios depend on this, while the strength of the products depends on the volume of binder created during the dissolution stage. However, this need is limited by the NaOH solubility in water, previous studies [31] show that a molar concentration higher than 24 is hard to achieve.

The compressive strength of GMT bricks, measured after 28 days of curing, exhibit a 35% increase, when the NaOH concentration was changed from 5 M to 15 M, also, an even higher effect (at 5 M the compressive strength increased from 1.4 MPa to 2.2 MPa, at 10 M it increases from 2.2 MPa to 3 MPa, while at 15 M NaOH the value increases from 3 to 3.5 MPa) was observed when the liquid to solid ratio was decreased from 0.25 to 0.15 [36].

Demir et al. [37] studied the possibility of obtaining GMT geopolymers for the removal of toxic and radioactive waste by realizing tests of Pb^{2+} adsorption from aqueous solutions. Firstly, the reaction degree depending on the curing temperature and Al_2O_3 addition was analyzed, accordingly, it was observed that higher Al_2O_3 content results in better reaction degree, while curing temperature showed minimum effects. Secondly, the Langmuir isotherm model and second-order kinetic model were identified as being the most suitable to evaluate the Pb^{2+} absorption. Accordingly, they state that the highest absorption efficiency obtained was 94% for GMT geopolymers with alumina addition.

In another study [38], the author evaluated the potential of obtaining geopolymeric concrete by replacing different percentages of type I Ordinary Portland Cement (OPC). The evaluation has been performed by compressive strength test on samples with 50% by wt. GMT (50GMT), 70%, wt. GMT (70GMT), 80% wt. GMT (80GMT) and 90% by wt. GMT (90GMT), respectively. According to their results, the presence of high percentages of tailing particles significantly increases the mechanical characteristics of the concrete, in 12 h cured samples at 70°C, despite the aging time (**Figure 12**).

Moreover, when the effects of harsh environment conditions (H_2SO_4, HNO_3, $MgSO_4$, Na_2SO_4 acids attack or high-temperature exposure) was evaluated, it was concluded that the increase of GMT also has good effects on these characteristics. However, if 55 days of immersion in acids reduce the compressive value to almost half, the high-temperature exposure at 1000°C almost destroyed the structure of the samples.

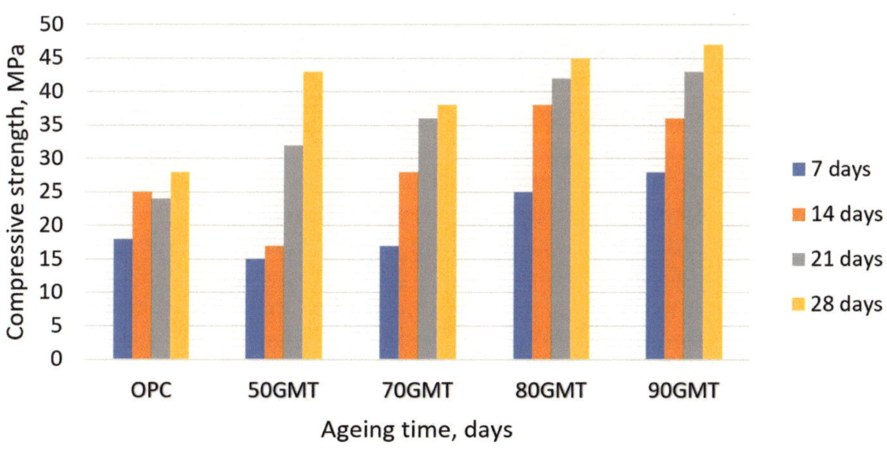

Figure 12.
Compressive strength evolution of OPC concrete with different percentages of GMT.

2.3.4 Phosphate mine tailings

The amount of tailings generated by a potassium mine depends primarily on the configuration of the potassium vein, the stability of the rocks and the mineral composition. All these are natural conditions that vary from one mine to another, from one warehouse to another and sometimes even within the same warehouse. Therefore, there is no standard model of mines in terms of processing and generation of finished products and tailings. Each mine has its specific conditions regarding the generation of liquid and solid tailings and their management. Thus, these specific conditions can vary throughout the life of a mine. However, for economic reasons, operators will seek to minimize the amount of mined tailing and processed [39].

Regarding the solid tailing facilities, waste management involves discharging it into the dump sites and refilling underground mines. The tailings from hot distillation and flotation, with sodium chloride as the main component, are dehydrated using centrifuges and filters, and then transported, employing conveyor belts, to the tailings dump. Besides, the dry process of electrostatic separation allows the dry management of the tailings at the tailings dump [40].

Regarding the liquid tailings, waste management involves discharge into groundwater (under certain geological conditions) and/or discharge into surface waters.

Fine Phosphate Sludge (PS) from the Moroccan Youssoufia plants were used for blended geopolymers obtaining by Moukannaa et al. [41]. According to their publication, PS rich in fluorapatite, quartz, calcite, dolomite, illite, palygorskite and hematite, needs supplementary aluminum content to present geopolymerisation potential, therefore, highly amorphous metakaolin was introduced in addition. Moreover, the reactivity increase was also performed by fusion activation at 550°C and 800°C, respectively. Based on the XRD analysis, it was observed that due to the presence of NaOH and high temperature, hematite decomposition occurs, while sodium aluminum silicate, sodium-calcium silicate, periclase and titanite formation is confirmed. Also, it was stated that the 550°C activations slightly reduce the content of dolomite, palygorskite, calcite and fluorapatite, while the quartz content seems to remain constant. However, at 800°C dolomite was greatly reduced, while the content of Na-rich phases significantly increases. On top of that, when the NaOH content was increased from 10–20%, the effects obtained at 800°C, were obtained at much lower fusion temperature (550°C). Surprisingly, based on the compressive strength results, the samples, cured at 28°C, made with raw materials fused at 550°C with 10% NaOH showed higher values than those with 20% NaOH or those fused at 800°C. Based on the microstructural analysis, it was stated that the poor strength, of 800°C fused samples, is related to the heterogeneous distribution of the sand particles, which acts as barriers for crack propagation in the matrix of 550°C fused samples (**Figure 13**). This phenomenon was also reported in previous studies [17, 42]. However, higher mechanical properties can be obtained by increasing the metakaolin content or by curing the samples at temperatures lower than 350°C.

The recycling possibility through geopolymerisation technology was also confirmed in [43]. According to the study, for optimum physical (low water absorption, high density etc.) and mechanical properties (compressive strength up to 62 MPa) a curing temperature of 83.33°C, a curing time of 14.50 days, and NaOH concentration of 12.5 M must be selected for phosphate sludge-based geopolymers mixed with fly ash or metakaolin.

2.3.5 Aluminum mine tailings (red mud)

Red Mud (RM) is the main secondary product (waste) that results from the Bayer process of refining bauxite to produce alumina. Generally, this process

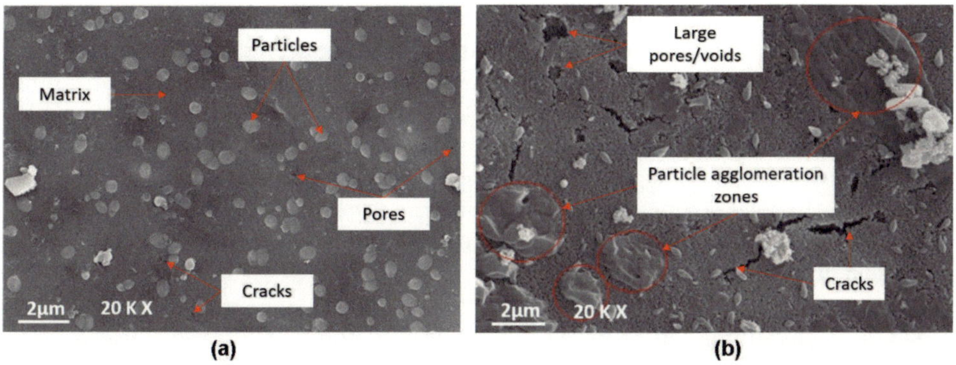

Figure 13.
Particle distribution effects on geopolymers microstructure: a) uniform distribution; b) heterogeneous distribution [41].

treats the aluminum ore (bauxite) with concentrated NaOH solutions at elevated temperature and pressure aiming to extract the gibbsite ($Al(OH)_3$), diaspore ($AlO(OH)$) and/or boehmite (γ-$AlO(OH)$). Therefore, RM is a strong alkaline waste due to the incomplete removal of NaOH. Accordingly, its composition depends on several factors but mainly on the composition of bauxite. The resulting amount of red mud also depends on the composition of the bauxite, however, it ranges from 0.33 up to 2 t of red mud per ton of alumina produced [5, 44]. Apart from the huge amount which is stored in tailing lakes, the main problem related to the environmental effects produced by this waste is its alkalinity. Moreover, considering that from Jan 1974 to Oct 2020 the global amount of extracted alumina was ≈2,814,091 thousand metric tonnes by multiplying with 1.1 (the average value of generated waste), the total amount of waste generated due to alumina extraction is ≈3,095,500 thousand metric tonnes [45]. According to the literature, only 2–3% of the produced waste is used in industrial-scale application, especially in civil engineering applications, therefore, almost 98% is being deposited [46].

Although, the methods of depositing and securing red mud facilities have been improved significantly over time, currently, worldwide more than 60 refineries extract alumina through the Bayer process depositing the waste in lakes [47, 48].

As can be seen from the literature, red mud is the most studied mine tailing for geopolymers obtaining [49, 50]. As stated in [49], Wagh et al. [51] published the most systematic study on the red mud use in alkali-activated materials. Accordingly, even from preliminary studies promising compressive strength (between 35 and 45 MPa) results have been obtained for RM as a single solid precursor. Moreover, by 15% metakaolin addition [52], RM becomes acts as a filler, and the compressive strength increase more than four times, compared with the geopolymer with RM as a single precursor. Depending on the replaced percentages, other authors obtained even better results [53, 54]. In terms of mechanical compressive, the addition of ground granulated blast furnace slag [55], rice husk ash [56] or fly ash [57] also showed significant improvements.

3. Conclusions

Based on the analyzed studies, the following conclusions can be drawn regarding the possibility to use mine tailings as precursors or components for geopolymers:

- Both mechanical and physical characteristics of geopolymers are affected by the water content introduced by the activator (NaOH concentration) and the curing parameters (time and temperature). Accordingly, higher NaOH concentrations and curing temperature results in an increase of the mine tailing reactivity.

- The introduction of aggregates in blended systems significantly influence the mechanical characteristics of geopolymers, generally, better properties are obtained. Moreover, their content can influence the dissolution degree of the matrix.

- The mine tailings can be used in multiple civil engineering applications since compressive strength up to 70 MPa can be achieved for 28 days cured samples.

- If appropriate manufacturing conditions (NaOH concentration, initial water content, forming pressure, and curing parameters) are selected, the copper mine tailings are feasible to produce eco-friendly bricks by a simple three steps method: (i) mixing the mine tailing with the activator, (ii) compress the mixture in a mold at a specific pressure, (iii) cure the products at slightly elevated temperature.

- Some types of mine tailings (especially sulphidic ones) show low interest as single precursors, generally, blast furnace slag additions are necessary to obtain suitable properties. Moreover, better results can be obtained by subjecting the mine tailing to mechanical grinding, calcination and alkali fusion.

- In corresponding conditions, gold mine tailing geopolymers are suitable for heavy metals encapsulation and immobilization.

- During the geopolymerisation, both, the crystalline and the amorphous phase from the tailings react in the alkaline environment, especially at high curing temperature. Also, higher temperature results in a denser structure.

- Thermal activation by drying or calcination removes the water content, creates many reactive phases and significantly increase the amorphous fraction of the aluminosilicate sources. Moreover, the alkaline fusion activation method significantly affects the mineralogical composition, resulting in Na-rich crystalline or amorphous phases.

- Mine tailings are available in enormous quantities worldwide (several thousand million tons per year) and can be used for cementitious application due to their high content of silicon and aluminum oxides content. Accordingly, by using them worldwide in the already developed applications and by encouraging their use in new ones, significant natural resources can be preserved.

- Five types of mine tailings have been identified and their potential use in geopolymerisation technology was evaluated.

- The main conclusion that can be drawn from this study is as follows, mine tailings can be successfully used for different application through geopolymerisation technology. The amount used currently is lower than the amount produced, therefore, the deposited quantities will increase. In order to avoid future disasters, related to tailings facilities failure, significant efforts must be made on the introduction of this waste in new, highly used goods.

Author details

Petrica Vizureanu*, Dumitru Doru Burduhos Nergis, Andrei Victor Sandu, Diana Petronela Burduhos Nergis and Madalina Simona Baltatu
"Gheorghe Asachi" Technical University of Iasi, Iasi, Romania

*Address all correspondence to: peviz2002@yahoo.com

References

[1] Davidovits, J. Geopolymers - Inorganic polymeric new materials. J. Therm. Anal. **1991**, *37*, 1633-1656, doi:10.1007/BF01912193.

[2] Geopolymer, Green Chemistry and Sustainable Development Solutions ... - Google Cărți Available online: https://books.google.ro/books?id=wIFo7L_zO8AC&pg=PA155&lpg=PA155&dq=kaoline+geopolymer+resistance+to+h2so4+davidovits&source=bl&ots=Fn2qsEaLi2&sig=ACfU3U0VjqlMZcEAsMY7PTDJWPGkQrBCvQ&hl=ro&sa=X&ved=2ahUKEwjawdLg2NjnAhUDuqQKHbC3Bu8Q6AEwA3oECAkQAQ#v=onepage&q=kaoline geopolymer resistance to h2so4 davidovits&f=false (accessed on Feb 17, 2020).

[3] Majidi, B. Geopolymer technology, from fundamentals to advanced applications: A review. Mater. Technol. **2009**, *24*, 79-87, doi:10.1179/175355509X449355.

[4] Duxson, P.; Fernández-Jiménez, A.; Provis, J.L.; Lukey, G.C.; Palomo, A.; Van Deventer, J.S.J. Geopolymer technology: The current state of the art. J. Mater. Sci. **2007**, *42*, 2917-2933, doi:10.1007/s10853-006-0637-z.

[5] Vizureanu, P.; Burduhos Nergis, D.D. *Green Materials Obtained by Geopolymerization for a Sustainable Future*; Materials Research Forum LLC, Ed.; Materials Research Foundations: 105 Springdale Lane, Millersville, PA 17551 U.S.A., 2020; Vol. 90; ISBN 978-1-64490-112-0.

[6] Hadi, M.N.S.; Zhang, H.; Parkinson, S. Optimum mix design of geopolymer pastes and concretes cured in ambient condition based on compressive strength, setting time and workability. J. Build. Eng. **2019**, *23*, doi:10.1016/j.jobe.2019.02.006.

[7] Zhuang, X.Y.; Chen, L.; Komarneni, S.; Zhou, C.H.; Tong, D.S.; Yang, H.M.; Yu, W.H.; Wang, H. Fly ash-based geopolymer: Clean production, properties and applications. J. Clean. Prod. 2016, *125*, 253-267.

[8] Van Dao, D.; Trinh, S.H.; Ly, H.B.; Pham, B.T. Prediction of compressive strength of geopolymer concrete using entirely steel slag aggregates: Novel hybrid artificial intelligence approaches. Appl. Sci. **2019**, *9*, doi:10.3390/app9061113.

[9] Vardhan, K.H.; Kumar, P.S.; Panda, R.C. A review on heavy metal pollution, toxicity and remedial measures: Current trends and future perspectives. J. Mol. Liq. 2019, *290*.

[10] Burduhos Nergis, D.D.; Vizureanu, P.; Ardelean, I.; Sandu, A.V.; Corbu, O.C.; Matei, E. Revealing the Influence of Microparticles on Geopolymers' Synthesis and Porosity. *Materials (Basel).* **2020**, *13*, 3211, doi:10.3390/ma13143211.

[11] Habert, G.; D'Espinose De Lacaillerie, J.B.; Roussel, N. An environmental evaluation of geopolymer based concrete production: Reviewing current research trends. *J. Clean. Prod.* **2011**, *19*, 1229-1238, doi:10.1016/j.jclepro.2011.03.012.

[12] Part, W.K.; Ramli, M.; Cheah, C.B. An Overview on the Influence of Various Factors on the Properties of Geopolymer Concrete Derived From Industrial Byproducts. In *Handbook of Low Carbon Concrete*; Elsevier Inc., 2016; pp. 263-334 ISBN 9780128045404.

[13] Wu, Y.; Lu, B.; Bai, T.; Wang, H.; Du, F.; Zhang, Y.; Cai, L.; Jiang, C.; Wang, W. Geopolymer, green alkali activated cementitious material: Synthesis, applications and challenges. Constr. Build. Mater. 2019, *224*, 930-949.

[14] Youssef, N.; Lafhaj, Z.; Chapiseau, C. Economic analysis of geopolymer brick manufacturing: A French case study. Sustain. **2020**, *12*, 7403, doi:10.3390/SU12187403.

[15] Nergis, D.D.B.; Abdullah, M.M.A.B.; Vizureanu, P.; Tahir, M.F.M. Geopolymers and Their Uses: Review. IOP Conf. Ser. Mater. Sci. Eng. **2018**, *374*, 12019, doi:10.1088/1757-899x/374/1/012019.

[16] Nergis, D.D.B.; Abdullah, M.M.A.B.; Sandu, A.V.; Vizureanu, P. XRD and TG-DTA study of new alkali activated materials based on fly ash with sand and glass powder. Materials (Basel). **2020**, *13(2)*, 343, doi:10.3390/ma13020343.

[17] Burduhos Nergis, D.D.; Vizureanu, P.; Corbu, O. Synthesis and characteristics of local fly ash based geopolymers mixed with natural aggregates. *Rev. Chim.* **2019**, *70(4)*, 1262-1267, doi: 10.37358/RC.19.4.7106.

[18] Gupta, M.; Kulkarni, N.H. A Review on the Recent Development of Ambient Cured Geopolymer Composites. In *International Conference on Emerging Trends in Engineering (ICETE)*; Springer, Cham, 2020; pp. 179-188.

[19] Nawaz, M. Introductory Chapter: Earth Crust - Origin, Structure, Composition and Evolution. In *Earth Crust*; IntechOpen, 2019.

[20] Hossain, S.K.S.; Roy, P.K. Fabrication of sustainable insulation refractory: Utilization of different wastes. Bol. la Soc. Esp. Ceram. y Vidr. **2019**, *58*, 115-125, doi:10.1016/j.bsecv.2018.09.002.

[21] Bouaissi, A.; Yuan Li, L.; Mustafa Al Bakri Abdullah, M.; Ahmad, R.; Abdul Razak, R.; Yahya, Z. Fly Ash as a Cementitious Material for Concrete. In *Zero-energy Buildings [Working Title]*; IntechOpen, 2020.

[22] Adiansyah, J.S.; Rosano, M.; Vink, S.; Keir, G. A framework for a sustainable approach to mine tailings management: Disposal strategies. J. Clean. Prod. **2015**, *108*, 1050-1062, doi:10.1016/j.jclepro.2015.07.139.

[23] Tailings.info • Conventional Impoundment Storage - The current techniques Available online: https://www.tailings.info/disposal/conventional.htm (accessed on Dec 12, 2020).

[24] Preventing Mine Tailings Disasters - Earthworks Available online: https://www.earthworks.org/campaigns/preventing-mine-waste-disasters/ (accessed on Dec 12, 2020).

[25] Ministry of Economy, R. Inventary and visual inspection of mine tailings facilities from Romania teritory Available online: http://www.economie.gov.ro/images/resurse-minerale/Raport Halde Iazuri 12 sept 2017.pdf (accessed on Dec 12, 2020).

[26] GRID-Arendal Global Tailings Dam Portal | GRID-Arendal Available online: https://www.grida.no/publications/472 (accessed on Dec 13, 2020).

[27] International Standard Organisation STRATEGIC BUSINESS PLAN ISO/TC 071 Available online: https://isotc.iso.org/livelink/livelink/fetch/2000/2122/687806/ISO_TC_071__Concrete__reinforced_concrete_and_pre-stressed_concrete_.pdf?nodeid=1162199&vernum=0 (accessed on Dec 13, 2020).

[28] *Common Recyclable Materials*; United States Environmental Protection Agency, 2015;

[29] Paiva, H.; Yliniemi, J.; Illikainen, M.; Rocha, F.; Ferreira, V. Mine Tailings Geopolymers as a Waste Management Solution for A More Sustainable Habitat. Sustainability **2019**, *11*, 995, doi:10.3390/su11040995.

[30] Obenaus-Emler, R.; Falah, M.; Illikainen, M. Assessment of mine tailings as precursors for alkali-activated materials for on-site applications. Constr. Build. Mater. **2020**, *246*, doi:10.1016/j.conbuildmat.2020.118470.

[31] Ahmari, S.; Zhang, L.; Zhang, J. Effects of activator type/concentration and curing temperature on alkali-activated binder based on copper mine tailings. J. Mater. Sci. **2012**, *47*, 5933-5945, doi:10.1007/s10853-012-6497-9.

[32] Ahmari, S.; Zhang, L. Production of eco-friendly bricks from copper mine tailings through geopolymerization. Constr. Build. Mater. **2012**, *29*, 323-331, doi:10.1016/j.conbuildmat.2011.10.048.

[33] Antonis, P.; Ioannis, P.; Maria, T. *Management of wastes from primary resource processing: identification, environmental evaluations*; 2017;

[34] Pacheco-Torgal, F.; Castro-Gomes, J.P.; Jalali, S. Investigations on mix design of tungsten mine waste geopolymeric binder. Constr. Build. Mater. **2008**, *22*, 1939-1949, doi:10.1016/j.conbuildmat.2007.07.015.

[35] Pacheco-Torgal, F.; Castro-Gomes, J.P.; Jalali, S. Adhesion characterization of tungsten mine waste geopolymeric binder. Influence of OPC concrete substrate surface treatment. *Constr. Build.* Mater. **2008**, *22*, 154-161, doi:10.1016/j.conbuildmat.2006.10.005.

[36] Kiventerä, J.; Golek, L.; Yliniemi, J.; Ferreira, V.; Deja, J.; Illikainen, M. Utilization of sulphidic tailings from gold mine as a raw material in geopolymerization. Int. J. Miner. Process. **2016**, *149*, 104-110, doi:10.1016/j.minpro.2016.02.012.

[37] Demir, F.; Derun, E.M. Modelling and optimization of gold mine tailings based geopolymer by using response surface method and its application in Pb2+ removal. J. Clean. Prod. **2019**, *237*, 117766, doi:10.1016/j.jclepro.2019.117766.

[38] Caballero, E.; Sánchez, W.; Ríos, C.A. Synthesis of geopolymers from alkaline activation of gold mining wastes. Ing. y Compet. **2014**, *16*.

[39] Procesul tehnologic in cadrul Proiectului Rosia Montana Available online: https://www.rmgc.ro/proiectul-rosia-montana/procesul-tehnologic-in-proiectul-minier-rosia-montana.html (accessed on Dec 17, 2020).

[40] INDUSTRY.GOV.AU | DFAT.GOV.AU *TAILINGS MANAGEMENT Leading Practice Sustainable Development Program for the Mining Industry*; 2016;

[41] Moukannaa, S.; Nazari, A.; Bagheri, A.; Loutou, M.; Sanjayan, J.G.; Hakkou, R. Alkaline fused phosphate mine tailings for geopolymer mortar synthesis: Thermal stability, mechanical and microstructural properties. J. Non. Cryst. Solids **2019**, *511*, 76-85, doi:10.1016/j.jnoncrysol.2018.12.031.

[42] Provis, J.L.; Duxson, P.; van Deventer, J.S.J. The role of particle technology in developing sustainable construction materials. Adv. Powder Technol. 2010, *21*, 2-7.

[43] Moukannaa, S.; Loutou, M.; Benzaazoua, M.; Vitola, L.; Alami, J.; Hakkou, R. Recycling of phosphate mine tailings for the production of geopolymers. J. Clean. Prod. **2018**, *185*, 891-903, doi:10.1016/j.jclepro.2018.03.094.

[44] Power, G.; Gräfe, M.; Klauber, C. Bauxite residue issues: I. Current management, disposal and storage practices. Hydrometallurgy **2011**, *108*, 33-45, doi:10.1016/j.hydromet.2011.02.006.

[45] World Aluminium — Alumina Production Available online: https://

www.world-aluminium.org/statistics/alumina-production/#data (accessed on Dec 18, 2020).

[46] Evans, K. The History, Challenges, and New Developments in the Management and Use of Bauxite Residue. J. Sustain. Metall. **2016**, *2*, 316-331, doi:10.1007/s40831-016-060-x.

[47] Essential Readings in Light Metals, Volume 1, Alumina and Bauxite - Google Cărți Available online: https://books.google.ro/books?id=1izXDQAAQBAJ&pg=PA466&lpg=PA466&dq=red+mud+is+twuice+in+quantity+compared+with+the+alumina+resulted+from+bayer+process&source=bl&ots=ydnFu1jckk&sig=ACfU3U0VQxfDfF_j1ZMn3Hw_xypbzR97nQ&hl=ro&sa=X&ved=2ahUKEwimt_eb8c7nAhUO2aQKHR40COsQ6AEwBHoECAoQAQ#v=onepage&q=red mud is twuice in quantity compared with the alumina resulted from bayer process&f=false (accessed on Feb 13, 2020).

[48] Khairul, M.A.; Zanganeh, J.; Moghtaderi, B. The composition, recycling and utilisation of Bayer red mud. Resour. Conserv. Recycl. 2019, *141*, 483-498.

[49] Hertel, T.; Pontikes, Y. Geopolymers, inorganic polymers, alkali-activated materials and hybrid binders from bauxite residue (red mud) – Putting things in perspective. J. Clean. Prod. 2020, *258*, 120610.

[50] Nuccetelli, C.; Pontikes, Y.; Leonardi, F.; Trevisi, R. New perspectives and issues arising from the introduction of (NORM) residues in building materials: A critical assessment on the radiological behaviour. Constr. Build. Mater. **2015**, *82*, 323-331, doi:10.1016/j.conbuildmat.2015.01.069.

[51] Wagh, A.S.; Douse, V.E. Silicate bonded unsintered ceramics of Bayer process waste. J. Mater. Res. **1991**, *6*, 1094-1102, doi:10.1557/JMR.1991.1094.

[52] Dimas, D.D.; Giannopoulou, I.P.; Panias, D. Utilization of alumina red mud for synthesis of inorganic polymeric materials. Miner. Process. Extr. Metall. Rev. **2009**, *30*, 211-239, doi:10.1080/08827500802498199.

[53] Xie, X.L.; Peng, P.; Zhu, W.F.; Wang, L.I. Preparation and fixation ability for Pb2+ of geopolymeric material based on bayer process red mud. In Proceedings of the Advanced Materials Research; Trans Tech Publications Ltd, 2014; Vol. 1049-1050, pp. 175-179.

[54] Bošković, I.; Vukcevic, M.; Vuk~evi}, M.; Turovi}, D.; Krgovi}, M.; Bo{kovi}, I.; Ivanovi}, M.; Zejak, R. Utilization of geopolymerization for obtaining construction materials based on red mud. *Mater. Tehnol.* **2013**, *47*, 99-104.

[55] Lemougna, P.N.; Wang, K.; Tang, Q.; Cui, X. Study on the development of inorganic polymers from red mud and slag system: Application in mortar and lightweight materials. Constr. Build. Mater. **2017**, *156*, 486-495, doi:10.1016/j.conbuildmat.2017.09.015.

[56] Geng, J.J.; Zhou, M.; Li, Y.; Chen, Y.; Han, Y.; Wan, S.; Zhou, X.; Hou, H. Comparison of red mud and coal gangue blended geopolymers synthesized through thermal activation and mechanical grinding preactivation. Constr. Build. Mater. **2017**, *153*, 185-192, doi:10.1016/j.conbuildmat.2017.07.045.

[57] Hu, W.; Nie, Q.; Huang, B.; Shu, X.; He, Q. Mechanical and microstructural characterization of geopolymers derived from red mud and fly ashes. J. Clean. Prod. **2018**, *186*, 799-806, doi:10.1016/j.jclepro.2018.03.086.

Chapter 3

Synthesis and Characterizations of High Carbon Ferrochrome (HCFC) Slag Based Geopolymer

Muktikanta Panigrahi, Saralasrita Mohanty, Ratan Indu Ganguly and Radha Raman Dash

Abstract

In the present Investigation, Geopolymer (GP) is made using High carbon Ferrochrome (HCFC) slag by synthesizing silicon and aluminum present in it with alkali liquid solution which binds other non-reactive materials present in the slag. The compressive strength of GP is found to be 11 MPa after 7-days of air curing. Increasing air curing time to 28-days, the strength is measured to be 15 MPa. XRD analyses indicate that there is gradual transformation of crystalline phases to non-crystalline glassy phases. SEM images show presence of more amount of glassy phase after air curing for a longer time. This is also corroborated with mechanical properties such as compressive strength. TGA results are also discussed for both uncured and cured GP samples. DSC isotherms indicate oozing out of inbuilt water present in the prepared GP materials which is an indication of condensation polymerization reaction occurring during the formation of Geopolymers.

Keywords: Geopolymer, Industrial waste, Slag, Cement, alkaline activator, phases, Absorption band Morphology, Compressive strength, Condensation polymerization

1. Introduction

Portland cement (PC) is a common structural material, which is widely used in construction of structure [1–4]. PC production causes environmental problems due to release of carbon dioxide (CO_2) [5]. The production of PC is extremely energy intensive process [6]. Therefore, research community have used industrial wastes such as fly ash (FA), pond ash (PA), slag, bottom ash (BA), quarry dust (QD), etc. as resource materials for the production of Geopolymer [1–5]. It is no wonder why researches are being carried out for the replacement of PC with new generation material, *i.e.*, GP. Therefore, GP technology has enabled to produce an alternative material to replace PC [7, 8].

One of the wastes is slag which is obtained from Ferro-alloy industries during the production of ferrochrome [1–5]. The slag contains variety of oxides such as magnesium oxide, iron oxide and chromium oxide which exist in more than one oxidation states. Also, Ferrochrome slag may contain elemental chromium [7, 8]. The major constituents of the slag are Cr_2O_3, FeO, SiO_2, MgO, Al_2O_3, CaO [7, 8]. In the year 2008, Davidovits has described GP to be a future constructional material and is predicted to replace PC in constructional areas [9, 10]. Reports are available which indicate waste materials *i.e.,* Fly Ash, Pond Ash, Bottom Ash, slag, etc. are suitable for production of Geopolymer [7, 8]. These oxides form chain-like structure by the reaction with active alkali solution such as sodium hydroxide (NaOH) or potassium hydroxide (KOH) in combination with sodium silicate [7, 8]. Thus, Geopolymer can be considered as an innovative constructional material for 21 century [11, 12]. Many researchers have described Geopolymer structure [7, 8]. The advantages of using GP materials lie in a fact that they are cheaper, unhazardous, fire resistant and environmental friendly [7, 8].

The corrosion test and p^H measurements have revealed that GP material has low leachability, excellent frictional resistance. Therefore, it can be used for construction of roads [13–16].

In the current work, HCFC slag is chosen from a ferrochrome industry as main component for preparing Geopolymer. Attempt is also made to characterize both HCFC slag and as-prepared Geopolymer material through different technique such as mechanical, structural, morphological, and thermal analyses.

2. Experimental details

2.1 Chemicals and materials

During the manufacture of Iron and Chrome alloy, slags are produced in smelter by the reaction of fluxing agents such as quartzite, bauxite, dolomite, corundum, lime and olivine with gangue materials of the ore. Usually, the foam slag in the smelting furnace contains 1.0–6.0 Cr depending on the charge at the reaction temperature.

In the present investigation, a HCFC slag from ferrochrome industry (**Shyam ferroalloys** Durgapur, West Bengal, India) is collected and composition is shown in **Table 1**. Other ingredients used are sodium hydroxide (NaOH) and sodium silicate (Na_2SiO_3), which are procured from M/S Loba Chemicals, India and Merck, India, respectively. These reagents are used for treating HCFC slag. NaOH acts as an activator and Na_2SiO_3 act as a polymerization agent. Water soluble plasticizer (SiKA) is used in the processing of Geopolymer which acts as a plasticizers *i.e.,* an agent for enhancing plasticity of the product (Geopolymer). The SiKA chemical is procured from a reputed firm located in the southern parts of India (Visakhapatnam market). All relevant details such as image, color, density, supplier name of raw materials used for preparing GP are described in **Table 2**.

2.1.1 Alkaline liquid

A combination of sodium silicate solution and sodium hydroxide solution are chosen as the alkaline liquid because they are cheaper [17–20].

Constituents of HCFC slag with percentages	SiO_2 = 28–31 MgO = 26.0–26.5 Al_2O_3 = 23.5–24.5 CaO =7.5–8.5 Cr_2O_3 = 4.5–5.4 FeO = 2.5–3.0
Sources	Syam Ferroallys, Durgapur

Table 1.
Chemical composition (%) of HCFC slag (XRF analysis).

Chemicals	Sample Images	Color	Density (g/mL)	Supplier
HCFC Slag		Blackish Gray	2.84 (Specific Gravity)	Shyam ferro alloys
Sodium Silicate		Brown Color	2.4 g/cm^3	Merck India
Sodium Hydroxide		Color less	2.13 g/cm^3	Merck India
Sika		Grayish	~1.18 kg/L (Relative density)	Vizag market

Table 2.
Ingredients sample image, color, density, supplier of HCFC slag.

2.1.2 Sodium hydroxide (NaOH)

The laboratory grade sodium hydroxide (NaOH) solid is taken in the form of pallets (3 mm, diameter), with a specific gravity of 2.130, 98% purity. NaOH solution is prepared by dissolving the pellets in distilled water. NaOH solution of 8 M concentration ($8 \times 40 = 320$ grams of NaOH solids per liter of the solution, where 40 is the molecular weight of NaOH) is used for preparing geopolymer. Similarly, NaOH solutions of other concentrations are prepared as discussed above.

2.1.3 Sodium silicate (Na_2SiO_3)

Usually, sodium silicate is common name of compounds which is expressed by a formula (Na2SiO2)nO. It is a semi-viscous solution having sp. gravity of 2.4 g/cm^3. The open chemical structure of sodium silicate is shown in **Figure 1**.

2.1.4 Plasticizer

In order to improve plasticity of the geopolymer, a SiKA compound is used along with reacting materials during processing of geopolymer. It is a low cost material. In addition, extra water (5–6 mL) is added to SiKA plasticizer solution. In designing

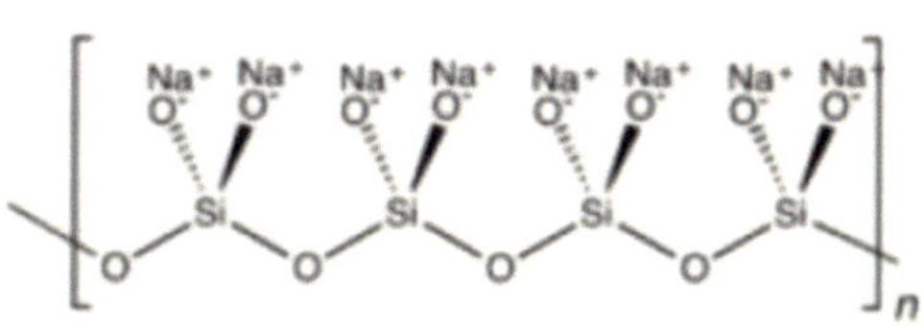

Figure 1.
Chemical structure of sodium silicate [Wikipedia.org/wiki/Sodium_metasilicate].

the mixtures the parameter "water-to-geopolymer solids ratio" by mass is considered to be important. Thus, the total mass of water is the sum of the mass of water contained in the sodium silicate solution, sodium hydroxide solution, and the mass of extra water [21, 22].

3. Synthesis procedure

Steps for geopolymer preparation with high carbon ferrochrome (HCFC) slag are described below (**Table 3**) [23, 24].

3.1 Raw material preparation

HCFC slag is collected from **Shyam ferroalloys** Durgapur, West Bengal, India in the form of lumps. The lumps are crushed and ground in a ball mill, followed by sieving through mesh (size 240). The fine particles, thus obtained, are dried in an oven for 2 hrs at 120°C.

The alkaline activator used for preparation of Geopolymer (GP) is a combination of sodium silicate and sodium hydroxide in appropriate proportion. While preparing the solution mixture in a flask, it is kept on a hot plate with a magnetic stirrer. The alkaline liquid is prepared by mixing sodium silicate solution and sodium hydroxide solution. The solution is continuously stirred with a magnetic stirrer.

Plasticizer (SiKA) with extra water is poured to the liquid mixtures.

3.2 Geopolymer preparation with raw materials

Prepared HCFC slag is poured into a mixture container. Then, alkali liquid solution is added slowly to the slag and is manually mixed. The mixture is left five minutes for polymerization reaction. Different mixture compositions are prepared.

3.3 Molding, casting and compaction

Geopolymer prepared with different variables are designated as S1, S2, and S3 (**Table 2**). The shape of mould is cylindrical 70 mm height and 55 mm diameter, respectively. The poured materials are rammed for compaction. The whole mixture is allowed to set for 3 minutes before demoulding samples. These are then rammed or struck for compaction followed by setting for 3 minutes before demoulding the sample. GP samples are removed from molds and are used for further investigations. It is to be mentioned their inside surface are coated with diesel oil to ensure smoother surfaces of samples. The entire process has been described (**Figure 2**).

Sample code	HCFC slag	Sod. Silicate (SS)	Alkali (8 M, NaOH) (SH)	Water soluble Plasticizer (Sika)
S1	85%	12%	3%	1-2 ml
S2	**85%**	**13%**	**2%**	**1-2 ml**
S3	85%	14%	1%	1-2 ml

Table 3.
Variation of SS and SH with other parameters as curing time 24 h, curing temperature 70 C, mesh 240 [24].

Figure 2.
Schematic diagram of process for preparation HCFC slag based GP product [23].

3.4 Curing

The green samples are kept for heating in an oven for 24 h at 70°C followed by curing in open air (7-days and 28-days) [25, 26].

Specimen(s) are used in different characterizations such as XRD, FTIR TGA, DSC, etc.

4. Characterization techniques

Different phases in the HCFC slag are identifies using X-ray diffraction technique. Essentially, the phases are determined by qualitative analysis of the phases through JCPD file. For this purpose, ferrochrome slags powdered are dried 110°C for 24 h and crushed into fine powdered use in the examination. For analysis purpose a Phillips PW-1710 Advance wide-angle X-ray diffractometer and Phillips PW-1729 X-ray generator using CuKα radiation (wavelength, λ = 0.154 nm). The generator is operated at 40 kV and 20 mA. The powder samples are placed on a quartz sample holder at room temperature and are scanned at diffraction angle 2θ from 10° to 60° at the scanning rate of 5°/min. X-rays diffraction studies are made with treated GP samples at different stages of curing. Similar analyses are also made with the as received HCFC slag for comparison purpose.

FTIR technique is used to analyze the presence of functional groups, formation of chemical linkage using a FTIR spectrometer (Nexus-870, Thermo Nicolet spectrophotometer) in range of 400–4000 cm^{-1}. The instrument parameters are kept constant (50 scan at 4 cm^{-1} resolution, absorbance mode). In this spectrometer, the IR radiations from an IR source are passed through the sample and the amount of energy adsorbed is recorded by suitable detector and is guided through an interferometer where a Fourier Transform is performed on the output signal. The pellet (13 mm diameter, 0.3 mm thick) so prepared is used for IR characterization. Before running the samples, a background spectrum is collected. Then pellets samples are put in a sample holder. The pellets are exposed to IR radiation in the spectrometer and data were collected.

Surface morphologies of prepared materials are analyzed by electron microscope (SEM using Carl Zeiss Supra 40). SEM is a microscope that uses electrons in place of light to produce image. In this measurement, the electron beam produced from electron gun is focused on a small portion of the sample that is kept in vacuum. Detector collects the output signals during the interaction of electrons with the sample and that is sent to a computer. This forms the final image. Special preparation technique is needed for the sample to avoid moisture absorption. All non-conducting materials need thin layer of conducting coating. This is done by 'sputter coater'. Operating voltage was 4 kV. The sample is placed in a small vacuum chamber. The argon gas is ionized in the applied electric field to form argon ion (Ar^+). The argon ions knock gold atoms from the surface of the gold foil and get deposited on sample. The elemental analyses are carried out with the help of EDX attached to the microscope.

TGA analysis is a type of thermal analysis that measures mass change of materials with change of temperature. The purpose of doing such experiment is to measure volatile content, thermal stability, degradation characteristics, etc. The as-prepared fine powdered of HCFC slag for Geopolymer preparation and as-prepared cured Geopolymer materials are also run in a thermogravimetric (TG) analyzer Perkin Elmer Pyris Diamond analyzer. Heating rate is taken 10°C/min during the experiments and is performed in an inert atmosphere (N_2 gas). The TGA run is carried out in the temperature range of 50–300°C.

Differential scanning calorimetry (DSC) is a technique for measuring the energy necessary to establish a nearly zero temperature difference between a substance and an inert reference material, as the two specimens are subjected to identical temperature regimes in an environment heated or cooled at a controlled rate. The technique provides qualitative and quantitative information about physical and chemical changes that involve endothermic or exothermic processes or changes in heat capacity using minimal amounts of sample. It has many advantages including fast analysis time, typically thirty minutes, easy sample preparation, applicability to both liquids and solids, a wide range of temperature applicability and excellent quantitative capability.

DSC has been used in the evaluation of small transitions such as multiple phase transitions in liquid crystals and those due to side chains in polymers which cannot be resolved by most other techniques. It allows accurate determination of temperatures associated with thermal events. Temperature can be calibrated with respect to one or more standards which allow highly accurate, precise and reproducible values. The technique reveals the thermal history imparted to thermoplastics as a result of different processing conditions. The information generated can be used to vary heating rates to deliver the required degree of crystallinity. Differential scanning calorimetry (DSC, Perkin Elmer Pyris Diamond) was used for determination of crystallization. Samples (5–10 mg) were placed in sealed aluminum pans and scanned under a constant nitrogen purge (20 mL/min). Subsequently, the samples were heated from 50–300°C at a rate of 10°C/min, cooled to 30°C at the same rate and held for 2 min to stabilize. Finally a second scan was carried out from room temp to 300°C. The result of enthalpy was noted from the second scan. DSC study is made for measurement of enthalpy of the two materials *i.e.*, as-received HCFC slag and GP prepared with HCFC slag. The test is carried out in nitrogen environment (N_2).

Tests to obtain the stress–strain curves prepared HCFC slag based GP in compression are performed using an Ultimate Tensile machine (UTM), AIMIL

COMPTEST 2000 model and is determined by standard ASTM which is inbuilt in the UTM machine. The tests on 100 × 200 mm GP cylinders are performed by using the displacement-control mode available in the test machine. It took approximately 30 to 60 minutes to complete each test in order to obtain the stress–strain curve(s). Loading in compression over a period is found to cause reduction in the measured value of the compressive strength of test cylinders. Maximum load is taken at the point of fracture of the sample. The specimen(s) is demoulded compression testing.

5. Results and discussion

Ferrochrome slag is obtained as waste material during production of ferrochrome alloy in a smelter. Pouring temperature of Slag is around 1500°C. The slag contains oxides of magnesium, iron and chromium in their different oxidation states. Chromium present in the slag may either be trivalent (+3 oxidation state) and or hexavalent (+6 oxidation state). Some elemental chromium may also be present in the slag. The oxides are usually spinel phase (*i.e.* $FeCr_2O_4$). $MgAl_2O_4$ spinel phase is also there in the slag. Presences of these spinel phases help the slag to enhance mechanical, chemical and thermal properties both at ambient and elevated temperatures [27, 28]. **Table 1** depicts chemical analysis of ferrochrome slag. It is evident that SiO_2, Al_2O_3, Fe_2O_3, and Cr_2O_3 are present in major quantities, while CaO, MgO, FeO are present in a minor quantities [29]. The ground and sieved (240 meshes) of as-received slag is dried for 2 h at 120°C for removal of moisture. The sieved HCFC slag is ready for Geopolymer preparation.

Geopolymer samples are made by treating with different molar concentration of sodium hydroxide (NaOH). Effect of different molar concentration of alkali materials on the compressive strength is studied. The strength properties of as-prepared Geopolymers are shown in **Figure 3**. S1, S2, S3, S4 represent Geopolymer prepared with 6 M, 8 M, 10 M, and 12 M solution of NaOH, respectively. These are cured at 70°C in an oven, followed by 7-days and 24-days of curing in air. It is well known that sodium plays an important role for polymerization of alumina (Al_2O_3) and silica (SiO_2), which is present in the slag. Each pair of bar drawn in **Figure 3** is

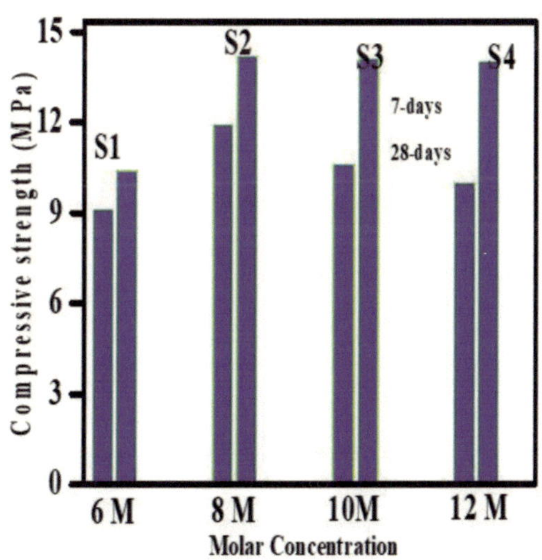

Figure 3.
Compressive strength vs. molar concentration at 7-days and 28-days air curing of HCFC slag based GP samples prepared by optimized parameters [23].

showing data on comparative strength scale for a Geopolymer prepared with different molar concentration, cured for 7 days and 28 days, respectively. It is found that maximum compressive strength has been achieved for the Geopolymer prepared with 8 M NaOH concentration [30–33].

Figure 4A and **B** show superimposed XRD pattern obtained from two different types of samples i.e., prepared with as-received HCFC slag (**Figure 4A**) and geopolymer prepared with optimal treatment combination (**Figure 4B**). Sharp XRD peaks can be seen in XRD pattern of as received slag (**Figure 4A**), whereas broaden peaks can be seen in geopolymer prepared with optimal treatment combination. It can infer that the slag material has crystalline phases and geopolymer has non-crystalline glassy phases. Sharp peaks in the as received material are identified to be quartzite and mulite phases, whereas broaden line profile of XRD pattern of geopolymer are glassy phases formed chemical and thermal treatment.

The morphology of HCFC slag and GP samples prepared with optimal treatment combination is studied (**Figure 5A** and **B**). **Figure 5A** shows SEM micrograph of HCFC slag which consists of fused mass of oxides with some glassy phase spread over the matrix. Since the slag material is poured from a high temp to a room temperature, therefore, faster cooling rate results. **Figure 5B** is a magnified image of Geopolymer as shown in the earlier slide. The picture shows morphology of Geopolymer prepared with optimal treatment combination. At lower magnification *i.e.,* 500 KX, it shows crystalline needle shape phase oriented in a random direction (**Figure 6B**). Additionally, there is some region of glassy phase co-exist together with the crystalline phase. The sample, therefore, is further investigated at a higher magnification *i.e.,* 1500 KX (**Figure 5B**). **Figure 5C** shows the bundles of rod-shaped phase oriented in the same direction. This crystalline phase has occurred due to cross-linking of alumino-silicate Geopolymer phase by reacting with the sodium of sodium hydroxide [34]. EDS analyses are done during the SEM studies (**Figure 7**). The elements such as Na, O, Al, Ca, Mg, Cr are present both in crystalline and glassy phases of as-prepared HCFC slag based GP (**Figure 7B**). However, all the elements present in GP are found in as-received slag except element such as sodium (**Figure 7A**). Occurrence of element sodium in GP is due to the reaction sodium compound with slag during polymerization process.

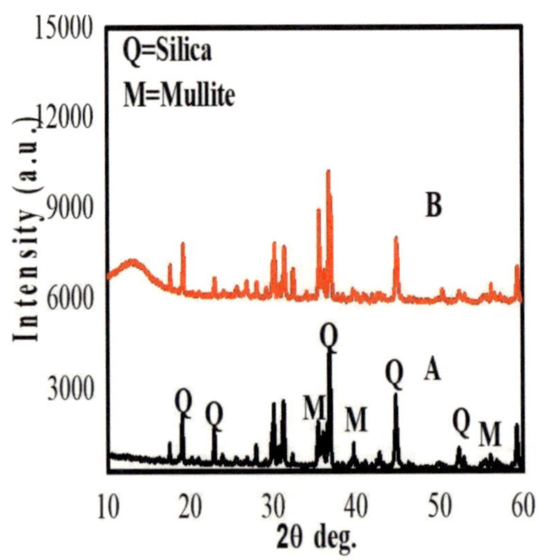

Figure 4.
XRD pattern of grinded with sieved as received HCFC slag (A) and as-prepared HCFC slag GP prepared by optimized parameters [23].

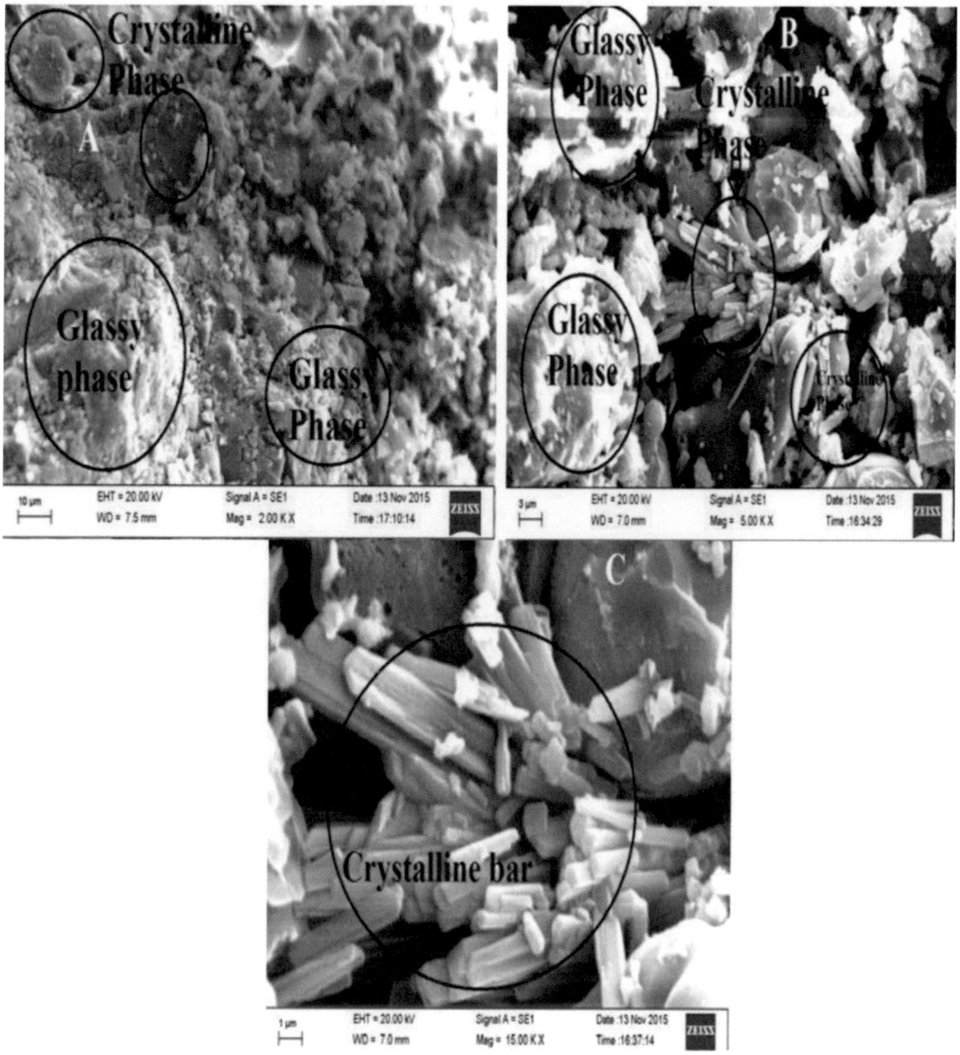

Figure 5.
SEM images of grinded with sieved of as-received HCFC slag (A), slag based GP samples prepared by optimized parameters [23].

FTIR spectra of the slag and GP material are shown in **Figure 6**. FTIR spectra indicate the presence of different absorption bands in each category of material. **Figure 6A** indicates the presence of different absorption bands occurring at different wave numbers *i.e.*, 3441, 2918, 1638, 1442, and 887 cm^{-1}. Bands are in agreement with stretching vibrations of O-H bonds (3441 cm^{-1} wave number) and H-O-H bending vibrations (1638 cm^{-1} wave number) of interlayer adsorbed H_2O molecule [35]. The hydroxyl-stretching band of water plays an important role and peak shift of the FTIR spectra is significant. Absorption band ensue at 887 cm^{-1} wave number is attributed to Si-O band and signifies the occurrence of silicate groups. Presences of $Al^{3+}O^{2-}$ absorption bands are also indicated near 805 cm^{-1} wave number [35]. Stretching bands is found at 440 cm^{-1} wave number and signifies the occurrence of Fe-O band [36]. However, the absorption bands are not found to occur in as-received HCFC slag. This is due to the formation of Geopolymer after treating and curing of as-received HCFC slag.

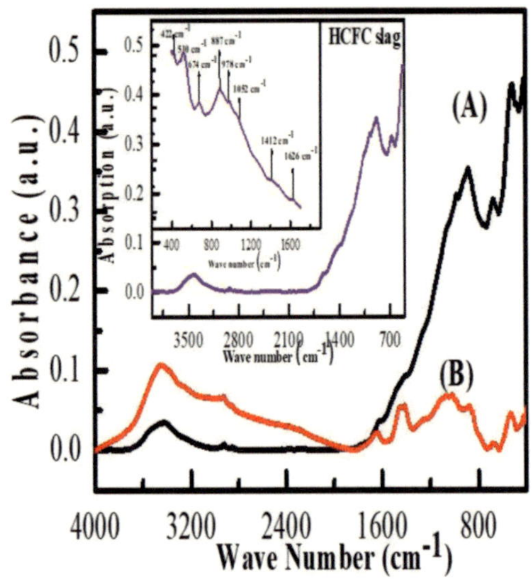

Figure 6.
FTIR spectra of grinded with sieved as-received HCFC slag (A) and HCFC slag based GP prepared from optimized parameters (B) [23].

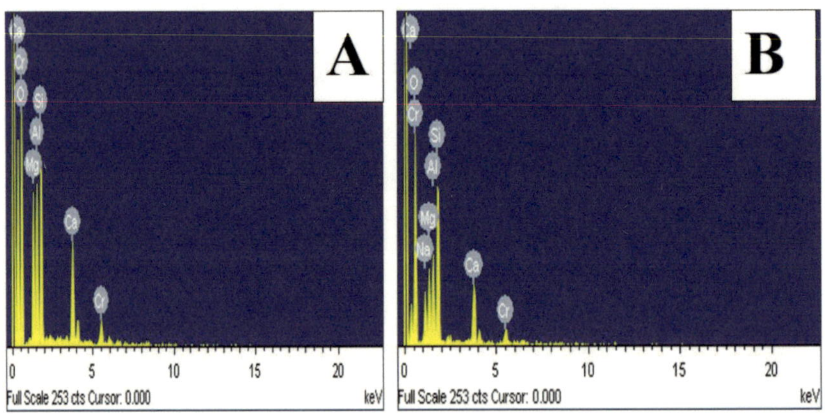

Figure 7.
EDS of grinded with sieved of as-received HCFC slag (A) and slag based GP samples prepared (B) by optimized parameters [23].

In TGA analysis, the weight loss is estimated for different temperature, ranging between 50 and 800°C and data are plotted as shown in **Figure 8**. **TGA runs** are taken from two samples, HCFC slag and GP. There is a sharp decrease in weight percentage beyond 100°C. This is attributed to the loss of water molecules. Between 100 to 350°C, there is a rapid declination the curve. Beyond 350°C, a little change is observed. The average total loss in weight (%) increases and decreases within the range of temperature 350–700°C [37, 38].

Differential scanning calorimetry (DSC) studies are made for both the samples i.e. HCFC slag and the GP formed it. It may be observed that there is no change in heat flow for slag (**Figure 9A**) whereas; there is significant change in heat flow for GP made from slag (**Figure 9B**) at 141°C. This is due to expulsion of physically bound water within the GP materials [39].

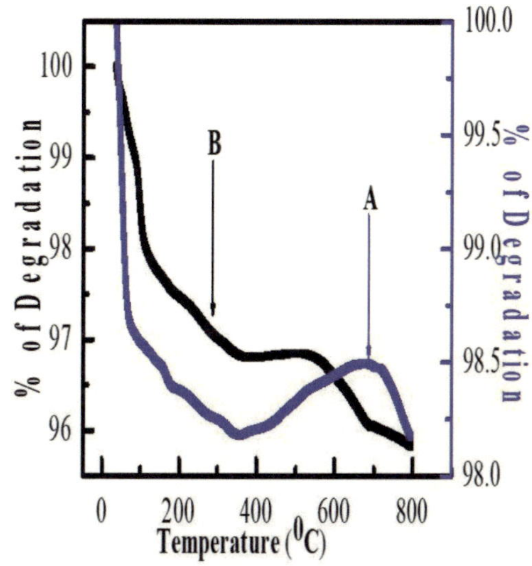

Figure 8.
TGA curve of grinded with sieved as-received HCFC slag (A) and HCFC slag based GP prepared with optimized parameters (B) [23].

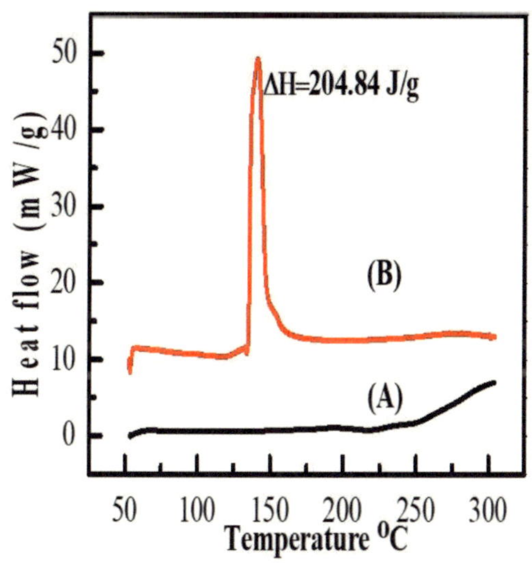

Figure 9.
DSC isotherm of grinded with sieved as-received HCFC slag (A) and HCFC slag based GP prepared with optimized parameters [23].

6. Conclusions

Geopolymers are prepared from HCFC slag successfully by treating with the alkali materials. The structure and morphology of Geopolymer are studied using different characterization technique. The microstructure of Geopolymer shows needle-shaped, randomly oriented crystalline phase, embedded in glassy phases. Together with microstructural observation, SEM/EDS analyses show presence of alkali materials in the Geopolymer which indicates occurrence of reaction of the HCFC slag with alkali solution during the Geopolymerization process. The maximum compressive strength is obtained to be 15 MPa by curing at 70°C in an oven for 24 h followed by cooling for 28-days in air. XRD pattern has clearly shown conversion of crystalline phase present in the slag has been transformed to glassy phase during the formation of Geopolymer. There is a marked difference observed in the FTIR spectrums. The numbers of peaks of as-prepared samples are much more in comparison to peaks present in as-received samples. This is due to formation of new bonds within the phases. TGA curves have revealed, for both the materials there is a gradual degradation material in the temperature range studied. DSC results show that there is a exothermic peak in Geopolymer at 150°C. This is due to elimination of water during condensation polymerization process occurring in the geopolymer.

Acknowledgements

The Authors gratefully acknowledge the **GIET University, Gunupur** for providing the facility to carry out the works. The authors are also acknowledging Central Research Facility (CRF), IIT Kharagpur, West Bengal, India.

Author details

Muktikanta Panigrahi[1*], Saralasrita Mohanty[2], Ratan Indu Ganguly[3] and Radha Raman Dash[4]

1 PG Department Materials Science, North Orissa University, Keonjhar, Odisha, India

2 School of Physical Sciences, National Institute of Science Education and Research (NISER), Jatani, Odisha, HBNI, India

3 Department of Metallurgical Engineering, National Institute of Technology (NIT), Rourkela, Odisha, India

4 National Metallurgical Laboratory (NML), Jamshedpur, Jharkhand, India

*Address all correspondence to: muktikanta2@gmail.com

References

[1] A. Bilodeau, V. Malhotra Mohan. High-volume Fly Ash System: Concrete Solution for Sustainable Development. **ACI Mater. J.**, Vol. 1, **2003,** pp. 41-48.

[2] L. Krishnan, S. Karthikeyan, S. Nathiya, K. Suganya. Geopolymer Concrete an Ecofriendly Construction Material. *Int. J. Research in Eng. and Tech.* (IJRET), **2014,** Vol. 3, pp. 2321-7308.

[3] Anthony Robin Pennell, Portland Cement Manufacture. **1978.** US4081285A.

[4] I. Afrani, C. Rogers. The Effects of Different Cementing Materials and Curing on Concrete Scaling. *Cement, Concrete, and Aggregates*, Vol. 16, **1994,** pp. 132-139.

[5] A. Allahverdi, E.N. Kani. Construction Wastes as Raw Materials for Geopolymers Binders. **Int. J. Civil Eng.,** Vol. 7, **2009**, pp. 154-160.

[6] E. Worrell, L. Price, N. Martin, C. Hendriks, L. O. Meida. Carbon Dioxide Emissions From The Global Cement Industry. **Annu. Rev. Energy Environ..** Vol. 26, **2001**, pp. 303-329.

[7] A.M.M. Al Bakri, H. Kamarudin, M. Bnhussain, I.K. Nizar, W.I.W. Mastura. Mechanism and Chemical Reaction of Fly Ash Geopolymer Cement- A Review. *Asian J. Scientific Research*, Vol. xxx, **2011.**

[8] J. Davidovits. Soft Mineralogy and Geopolymers. *Proc. 1st Int. Conf. on Geopolymers*, Compiegne, France, Vol. 1, **1988, pp.** 19-23.

[9] Z. Zhang, J.L. Provis, A. Reid, H. Wang. Review Geopolymer Foam Concrete: An Emerging Material for Sustainable Construction. *Construct. and Building Mater.* Vol. 56, **2014,** pp. 113-127.

[10] J. Davidovits. Geopolymer Chemistry and Applications. Edition: 2nd (**2008**), 3rd (**2011**), 4th (**2015**), 5th (**2020**), Publisher: Institute Geopolymer, Geopolymer Institute, Saint-Quentin, France, Editor: Joseph Davidovits, ISBN: 9782954453118 (5th ed.)

[11] D.D. Sakthidoss, T. Senniappan. A Study on High Strength Geopolymer Concrete with Alumina-Silica Materials Using Manufacturing Sand. **Silicon,** Vol. 12, **2020,** pp. 735-746.

[12] M. Panizza, M. Natali, E. Garbin, S. Tamburini, M. Secco. Assessment of Geopolymers with Construction and Demolition Waste (CDW) Aggregates as a Building Material. **Construction and Building Materials,** Vol. 181, **2018**, pp. 119-133.

[13] B.B. Lind, A.-M. Fallman, L.B. Larsson. Environmental Impacts of Ferrochrome Slag in Road Construction. **Waste Management**, Vol. 21, **2001,** pp. 255-264.

[14] W. Shao-Peng, Y. Wen-Feng, X. Yong-Jie, L. Zhen-Hua. Design and Preparation of Steel Slag. **J. Wuhan University of Technology**, Vol.18, **2003**, pp. xxx-xxx.

[15] M. Tossavainen. Leaching Results in the Assessment of Slag and Rock Materials as Construction Material. **2005**. (http://epubl.ltu.se/1402-1544/2005/44/LTU-DT-0544-SE.pdf - accessed on 31 August 2018)

[16] M. Kauppi, N. Pekka. Production, Characteristics and Use of Ferrochrome Slags. **2007**, Infacom 7.

[17] S.V. Patankar, Y.M. Ghugal, S.S. Jamkar. Effect of Concentration of Sodium Hydroxide and Degree of Heat Curing on Fly Ash-Based Geopolymer Mortar. **Indian J. Mat. Sci.,** Vol. 2014, **2014**. Pp. 6-xxx.

[18] S. Kumar, P.D. Gautam, B.S. Chandra Kumar. Effect of Alkali Activator Ratio on Mechanical Properties of GGBS based Geopolymer Concrete. **Int. J. Innovative Techn. & Exploring Engg.** (IJITEE), Vol.8, **2019**, pp. xxx-xxx.

[19] A.B. Malkawi, M.F. Nuruddin,' A.F, H. Almattarneh, B.S. .Mohammed, Effects of Alkaline Solution on Properties of the HCFA Geopolymer Mortars, *Procedia Engineering*, Vol. 148, **2016**, Pp. 710-717.

[20] S.H. Sanni, Performance of Alkaline Solutions on Grades of Geopolymer Concrete. **Int. J. Research in Engg. Techn.**, Vol. 02, **2013**, pp. 366-371.

[21] R.R. Reddy Bellum, R. Nerella, S.R. Chand Madduru, C.S. Reddy Indukuri, Mix Design and Mechanical Properties of Fly Ash and GGBFS-Synthesized Alkali-Activated Concrete (AAC). **Infrastructures**, Vol. *4, 2019* , pp. 20-xxx.

[22] A. Motorwala, V. Shah, R. Kammula, P. Nannapaneni, P.D.B. Raijiwala. ALKALI Activated FLY-ASH Based Geopolymer Concrete. **J. Artic.**, Vol. 3, **2013**, pp. 159-166.

[23] M. Panigrahi, S. Mohanty, R.I. Ganguly, R.R. Dash, An advanced cured high carbon ferrochrome slag (HCFCS) geopolymer (GP): A constructional materials, **IOP Conf. Series: Materials Science and Engineering,** Vol. 410, **2018**, pp. 012002-012xxx.

[24] M. Panigrahi, S. Mohanty, R.R. Dash, R.I. Ganguly, Development of Novel Constructional Material from Industrial Solid Waste as Geopolymer. **IOP Conf. Series: Materials Science and Engineering**, Vol. 410, **2018**, pp. 012012-012xxx.

[25] A.A. Patil, H.S. Chore, P.A. Dodeb. Effect of Curing Condition on Strength of Geopolymer Concrete. **Advances in Concrete Construction**, Vol. 2, **2014**, pp. 29-37.

[26] M.F. Nurruddin, S. Haruna, B.S. Mohammed, I.G. Sha'aban, Methods of Curing Geopolymer Concrete: A Review, **Int. J. Adv. Applied Sci.**, Vol. 5, **2018**, pp. 31-36.

[27] C.N. Harman, N.S.S. Rama Rao. A Process for the Recovery of Chromium and Iron Oxide in high Carbon Ferro Chrome Slag to Obtain Chromium and Iron in the form of Saleable Metal. The thirteenth International Ferroalloys Congress Efficient technologies in ferroalloy industry, Almaty, Kazakhstan. **2013**, pp. 103-108.

[28] H. Wanga, M.M. Nzottab, L. Tengc, S. Seetharamanc. Decarburization of Ferrochrome and High Alloy Steels with Optimized Gas and Slag Phases towards Improved Cr Retention. *J. Min. Metall. Sect. B-Metall.,* Vol. 49, B, **2013**, pp. 175-181.

[29] M. Erdem, H.S. Altundogan, M.D. Turan, F. Tumen. Hexavalent Chromium Removal by Ferrochromium Slag. **J. Hazardous Mater. B**, Vol. 126, **2005**, pp. 176-182.

[30] Z. Zhang, J.L. Provis, A. Reid, H. Wang. Review Geopolymer Foam Concrete: An Emerging Material for Sustainable Construction. **Construct. and Building Mater.**, Vol. 56, **2014**, *pp.* 113-127.

[31] M.I. Khan, K. Azizli, S. Sufian, A.S. Khan, H. Ullah, Z. Man. Geopolymer as Sustainable Binder of the 21st Century: A Review. *World Sustainability Forum*, **2014**-Conf. Proceedings Paper.

[32] V. Kumar, V.K. Singh, A. Srivastava. Low Temperature Synthesis of High Alumina Cements by Novel Co-Melt Precursors and Their Implementation as Castables with Some Micro Fine Additives. **J. Am. Ceram. Soc.**, Vol. 96, **2013**, pp. 2124-2131.

[33] P.H. Kumar, A. Srivastava, V. Kumar, M.R. Majhi, V.K. Singh. Implementation of Industrial Waste Ferrochrome Slag in Conventional and Low Cement Castables: Effect of Micro Silica Addition. **J. Asian Ceramic Soc.**, Vol. *2*, **2014,** pp. 169-175.

[34] S.D. Muduli, J.K. Sarangi, B.D. Nayak, B.K. Mishra. Effect of NaOH Concentration in Manufacture of Geopolymer Fly Ash Building Brick. **Greener J. Phys. Sci.**, Vol. 3, **2013**, pp. 204-211.

[35] J.D. Russell, Infrared methods. In: A Hand Book of Determinative Methods in Clay Mineralogy. *Wilson MJ (Ed.) Blackie* **and Son Ltd**, NY, Vol. 133, **1987,** pp. 11-67.

[36] K. Ramaswamy, R. Venkatachalapathy,. Infrared Spectroscopic Analysis of Sedimentary Formations of Neyveli Lignite Minecut-Ii. **Indian J. Pure Appl. Phys**., Vol. 30, **1992**, pp. 171.

[37] X. Gao, Q.L. Yu, H.J.H. Brouwers. Reaction Kinetics, Gel Character and Strength of Ambient Temperature Cured Alkali Activated Slag–Fly Ash Blends. *Construct. & Building Mat.,* Vol. 80, **2015,** pp. 105-115.

[38] F. Škvára, L. Kopecký, J. Nìmeèek, Z. Bittnar, Microstructure of Geopolymer Materials Based on Fly Ash. **Ceramics –Silikáty,** Vol. 50, **2006**, pp. 208-215.

[39] S. Alehyen, M. EL Achouri, M. Taibi, Characterization, Microstructure and Properties of fly ash-based Geopolymer. *J. Mat. Environ. Sci.,* Volume 8, **2017,** pp 1783-1796.

Chapter 4

Geopolymer Concrete under Ambient Curing

M. Indhumathi Anbarasan, S.R. Sanjaiyan and S. Nagan Soundarapandiyan

Abstract

Geopolymer concrete (GPC) has significant potential as a more sustainable alternative for ordinary Portland cement concrete. GPC had been introduced to reduce carbon footprints and thereby safeguarding environment. This emerging eco friendly construction product finds majority of its application in precast and prefabricated structures due to the special curing conditions required. Sustained research efforts are being taken to make the product suitable for in situ applications. The developed technology will certainly address the issues of huge energy consumption as well reduce water use which is becoming scarce nowadays. Ground Granulated Blast Furnace Slag (GGBS) a by-product of iron industries in combination with fly ash has proved to give enhanced strength, durability as well reduced setting time. This study investigates the effect of GGBS as partial replacement of fly ash in the manufacture of GPC. Cube and cylindrical specimens were cast and subjected to ambient curing as well to alternate wetting-drying cycles. The 28 day compressive strength, split tensile strength, flexural strength and density of GPC specimens were found. The study revealed increase in compressive strength, split tensile strength, density as well flexural strength up to 40 percent replacement of fly ash by GGBS.

Keywords: geopolymer concrete, ambient curing, Flyash, GGBS, Setting time

1. Introduction

It is well known that concrete is one of the most widely used reliable and effective construction material all over the World [1, 2]. Rapid urban growth development leads to the usage of concrete in construction industry increase day by day which further increases the demand of ordinary Portland cement. In order to meet the huge demand, the production of ordinary Portland cement (OPC) increases every year. During the production of OPC, an enormous amount of greenhouse gas such as carbon dioxide (CO_2) will be emitted giving rise to global warming issues [3]. As such invention of alternative binding materials evolved. Geopolymer concrete proved itself to address the issues by reduced carbon footprints. Further, the enormous energy required in the production of OPC could be avoided leading to energy conservation.

Geopolymer had proved further superior to OPC in terms of acid resistance, sulphate resistance, withstanding heat [4], fire and possessing corrosion resistance. Geopolymerization involves a heterogeneous chemical reaction between silicon and aluminium in a source of geological origin or industrial by products such as flyash and GGBS (binder) with high alkaline solution of sodium hydroxide and sodium silicate resulting in the form of three-dimensional amorphous to semi crystalline polymeric and ring structure comprising Si-O-Al and Si-O-Si bonds.

There exist some limitations in practical usage of geopolymer concrete in construction industry. Usage of alkaline solution in geopolymer concrete contributes to cost aspect. Further, requirement of specialized curing namely heat curing or steam curing makes it difficult for in situ applications.

Due to the heavy need of electricity, numerous thermal power stations have been installed throughout the country which gives rise to fly ash generation. Flyash is a heterogeneous by-product material produced in the combustion process of coal used in power stations. Fly ash particles are almost spherical in shape which allows them to flow and blend freely in mixtures. This characteristic makes fly ash a desirable binder for concrete. Further, control of high thermal gradients, pore refinement, depletion of cement alkalis, resistance to chloride [5] and sulphate penetration and continued micro structural development through a long-term hydration and pozzolanic reaction contributes to added durability aspects. Also, the magnitude of reinforcement steel production is also enormous so as to meet the present day needs of multistoreyed structures. The by product of iron manufacturing by heating iron ore, lime stone and coke at very high temperature of 1500 degree celcisus is GGBS.

GGBS posseses good mobility characteristics arising from consistent fineness, unique particle shape and from lower relative density. Also, workability gets improved due to the smoother surface texture and glassy surface of GGBS. Also, efflorescence and staining of concrete shall be prevented by the use of GGBS. It is evident that due to its superior performance, it replaces Sulphate Resisting Portland Cement (SRPC) and useful against severe chloride attack in reinforced concrete in marine environment.

In this work, an attempt was made to study the performance of GGBS admixed geopolymer concrete under ambient curing condition. The optimum percentage replacement of fly ash by GGBS had been determined by conducting compressive strength and split tensile strength tests. The results obtained would help resolve the issues addressed above.

2. Experimental procedure

2.1 Materials

In this study, Class F fly ash with specific gravity 2.3 and GGBS having specific gravity 2.1 was used. River sand was used as fine aggregate as per standards [6].

Coarse aggregate of different sizes 7 mm, 14 mm and 20 mm was graded using sieve analyser with specific gravity 2.717, 2.81, 2.76 has determined using pycnometer. The water absorption was found as 0.26%, 0.15%, 0.42% respectively while the aggregates are immersed in water for 24 hours. Sodium hydroxide (NaOH) having 14 M molarity with 97–98% purity in the form of pellets was used [7]. Sodium silicate available in the form of gel was used. Both chemicals have brought from the local supplier (Thirumala Nadar and Sons, Madurai, India-625001).

2.2 Concrete mix proportions and specimen preparation

The geopolymer concrete comprises binder, fine aggregate, coarse aggregate and alkaline solution. The materials were quantified before mixing using standard guidelines as per Rangan and Hardjito method [8]. Sodium hydroxide was mixed with water and then sodium silicate was added to the diluted sodium hydroxide to make alkaline solution. The alkaline solution was prepared one day before concrete mix to facilitate polymerization process.

The percentage replacement of fly ash by GGBS was varied from 0–50%. Including the control mix, six mix proportions were used for both ambient curing and for wetting-drying cycles. Due to the reason that quick setting developed [9], percentage replacement was not done above 50%.

It is a usual procedure to first mix the binder materials, fine aggregate and coarse aggregate using mixer machine [10] for about three to four minutes. Then the prepared alkaline liquid should be poured over the dry mix concrete and mixing to be continued for about four to five minutes to initiate geopolymerisation process.

After getting an intimate mix, they were cast using moulds. The cast specimens were cured under ambient curing for 28 days while another set of specimens were subjected to wetting for 2 days in curing tank and drying for 2 days in ambient condition. These wetting and drying cycles were continued upto 28 days. The cast cube and cylinder specimens subjected to ambient curing are shown in **Figures 1** and **2** respectively.

Figure 1.
Cube specimens under ambient curing.

Figure 2.
Cylinder specimens under ambient curing.

Prisms of size 500 mm x 100 mm x100mm were cast to study the flexural behaviour of GPC using fly ash and GGBS combination. Prisms exposed to ambient curing for 28 days are shown in **Figure 3**.

The nomenclature of GPC specimens is presented in **Table 1**.

2.3 Test methods

To ascertain mechanical behavior of geopolymer concrete under ambient curing, compressive strength, split tensile strength and flexural strength tests were carried out. SEM analysis was done to study the microstructure, particles and formation of specimen [11].

2.3.1 Compressive strength test

Cube specimens were subjected to compressive axial load applied at the rate of 1.2 N/mm^2 to 2.4 N/mm^2 in accordance with IS code 5816:1999 [12] (**Figure 4**). The load was applied in such a way that the fracture plane crosses the trowel surface.

2.3.2 Split tensile strength test

To measure the tensile strength of concrete, cylinder specimens are placed centrally between the plates and the load is gradually applied in such a way that fracture plane will pass along its vertical diameter of specimen (**Figure 5**). The test was carried out as per IS code 5816:1999 [12] speficications.

Figure 3.
Ambient cured prism specimens.

Specimen (Indication)	Description
G	GGBS
0,10.20,30, 40,50	Percentage of GGBS in Fly ash
A	Ambient curing
A'	Wetting Drying Cycles

Table 1.
Nomenclature of GPC specimens.

Figure 4.
Compressive strength test of GPC.

Figure 5.
Split tensile strength test of GPC.

2.3.3 Flexural strength test

The flexural strength test set up is shown in **Figure 6**. Load is applied to the uppermost surface as cast along two lines 150 mm apart. The load was increased continuously till the specimen failed recording the failure load.

2.3.4 Microstructural analysis

To ascertain the morphological features of geopolymer concrete made with new combination of fine aggregates, micro structural analysis becomes essential.

Figure 6.
Flexural strength test of Prisms.

The test results would reveal the factors affecting the mechanical properties. Presently available micro structural analysis include Scanning Electron Microscope (SEM), X Ray Diffraction Analysis (XRD), Energy Dispersive Spectroscopy (EDS) or Energy Dispersive X-Ray Analysis (EDAX) etc. In the present investigation, SEM and EDAX were used for the purpose for all mixes.

2.3.4.1 Energy dispersive X-ray analysis (EDAX)

EDAX shall very well be utilised to obtain the elemental composition or chemical characterisation of an area of interest on a specimen. The EDX or EDAX spectrum displays spectra showing peaks corresponding to the energy levels of elements making up the true composition of samples being analysed as received by X-rays. An excitation source, X-ray detector, pulse processor and the analyser constitute the primary components of EDAX. It is also possible to get elemental mapping of a sample as well image analysis. Samples of all proportions were made ready in powder form for analysis. For better accuracy, quantitative correction procedures are need to be applied.

2.3.4.2 Scanning electron microscopy (SEM)

In SEM, the surface of specimens will be scanned with a focused beam of electrons and images would be produced. The atoms in the sample will be interacted by the electrons and produce signals giving information regarding the topography of surface and composition of the sample. Resolution in the range of even 1 nanometer shall be achieved with the help of SEM.

A scanning electron microscope within built energy dispersive X-ray analysis (SEM- EDX) serves as good supplement to the optical microscope in examining new, old and deteriorated concrete. SEM analysis was done both for the samples having combination of fine aggregates (DMS and M-Sand) as well for samples with DMS alone. Samples for SEM analysis were collected from destructed pieces.

Thus, various tests as discussed above were conducted on geopolymer mortar as well on geopolymer concrete specimens conforming to relevant codal provisions.

3. Results and discussion

3.1 Compressive strength

The obtained compressive strength values for mixes G10A, G20A, G30A, G40A, G50A and G0A are respectively 30.88, 34.22, 37.78, 41.78, 36.22 and 26.5 MPa and the compressive strength values for mixes G10A', G20A', G30A', G40A', G50A'and G0A' are 30.56, 33.85, 36.65, 41.52, 38.5 and 27.76 MPa respectively. The values are plotted in **Figures** 7 and **8**.

The test results yield an inference that addition of GGBS in fly ash increases the compressive strength upto 40% replacement [13, 14]. The presence of calcium content in GGBS seems to be a real contributor for increase in strength upto 40% after which microstructure gets altered causing decrease in compressive strength. The compressive strength of control mix was low compared to other mixes using GGBS.

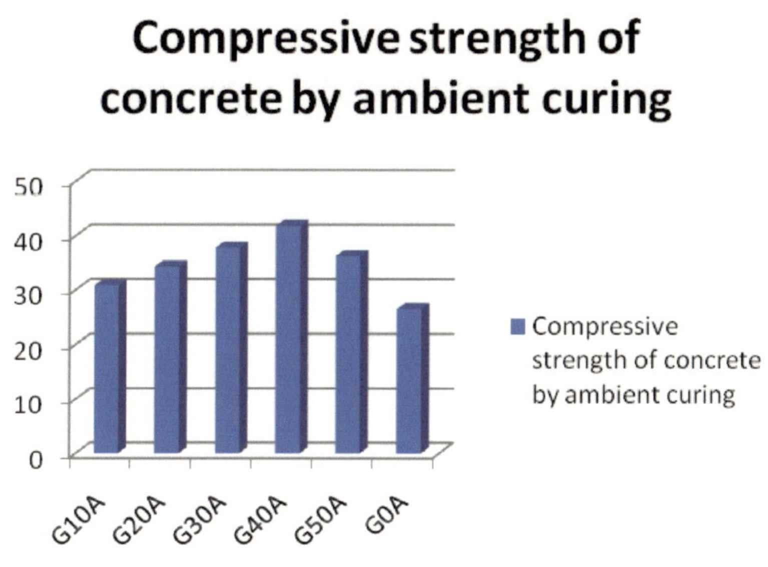

Figure 7.
Comparison of Compressive strength of GPC by ambient curing.

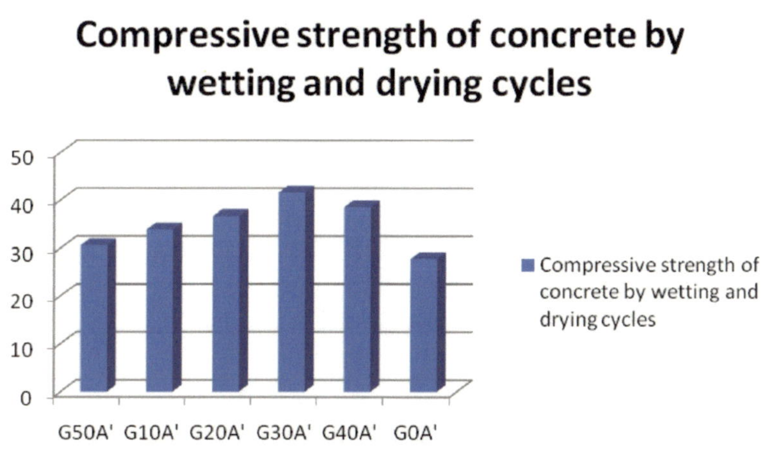

Figure 8.
Comparison of Compressive strength of GPC by Wetting and drying cycles.

3.2 Split tensile strength

The split tensile strength after 28 days for mixes G10A, G20A, G30A, G40A, G50A and G0A are 3.54, 3.96, 4.24, 4.81, 4.05 and 2.86 MPa respectively and these values are represented in **Figure 9**. The test results indicate that the maximum split tensile strength of 4.81 MPa is achieved for G40A mix. This may be due to the fact that addition of GGBS upto 40% increases the consistency, mortar phase and bond strength between mortar and aggregate. However, the split tensile strength of concrete decreased after 40% addition of GGBS which may be due to the glossy surface, surface texture and relatively low density reducing the bond strength and consistency (**Figure 10**).

3.3 Density

The measured density for the mixes G10A, G20A, G30A, G40A, G50A and G0A are respectively 2267, 2306, 2378, 2392, 2384 and 2240 kg/m^3 as shown in **Figure 10**.

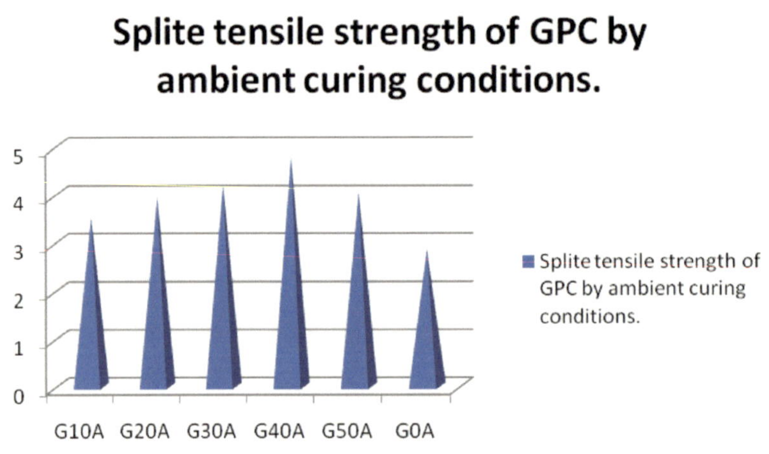

Figure 9.
Split tensile strength of GPC all combinations.

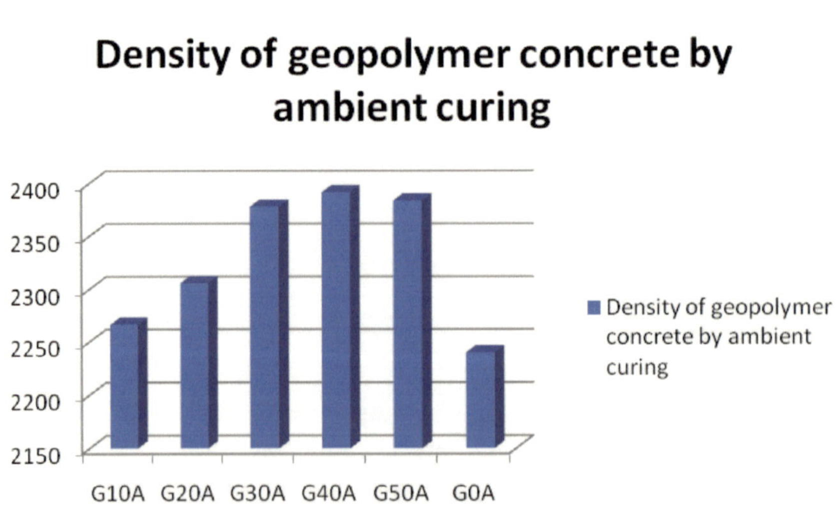

Figure 10.
Density of GPC for all combinations by ambient curing.

and the density values for mixes G10A', G20A', G30A', G40A', G50A' and G0A are 2254, 2258, 2327, 2385, 2346 and 2217 kg/m^3 as shown in **Figure 11**. Here again, upto 40% replacement, increase in density could be noticed.

3.4 Flexural strength

The flexural strength values of mixes G10A, G20A, G30A, G40A, G50A and G0A are 2.35, 3.59, 4.39, 6.5, 5.84 and 5.35 MPa respectively as shown in **Figure 12**. Flexural strength test also yields the same inference that increase in strength upto 40% replacement of flyash by GGBS and getting decreased after that.

3.5 Microstructural analysis

EDAX and SEM analysis were carried out on specimens as a part of microstructural analysis [15] to know the factors contributing to the behaviour of new combination of fine aggregates used in geopolymer concrete.

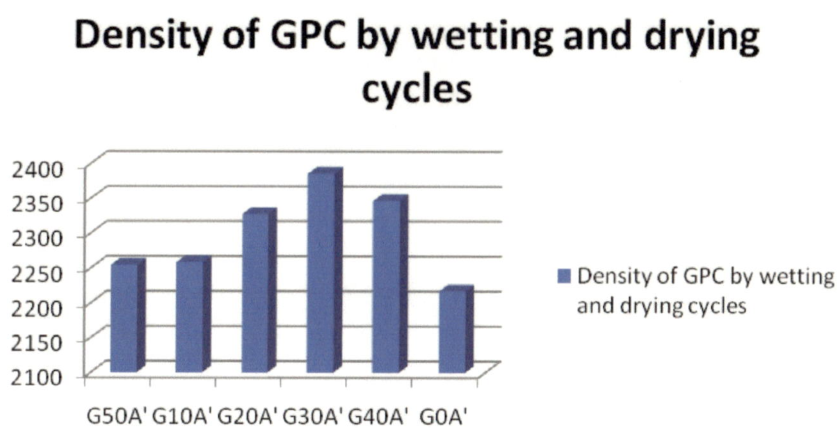

Figure 11.
Density of GPC for all combinations by wetting and drying cycles.

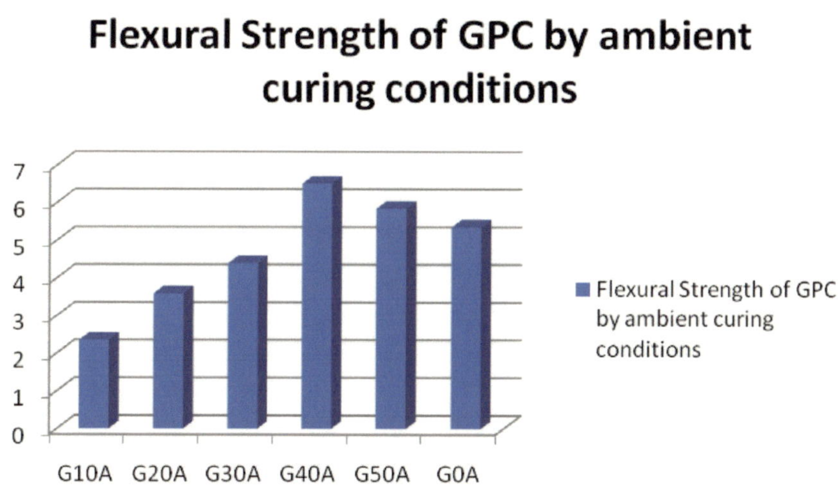

Figure 12.
Flexural strength of GPC under ambient curing conditions.

3.5.1 EDAX analysis

The elemental composition of D20, D40, D60, D80 and D100 specimens are shown in **Figures 13–17** which show the presence of Ca, Si, Al confirming the formation of geopolymeric gel as well the process of geopolymerisation. The presence of calcium in geopolymer concrete indicates that the setting time fastens during process. The rapid hardening process through solidification of gel frame work is found to be contributed by C-A-S-H gel. This would facilititate ambient curing conditions.

Elemental Weights obtained from EDAX analysis are given in **Table 2**. It shall be observed that oxides are present in larger extent in all mixes. Binding property could be weekened by the presence of oxides. The decrease in strength initially (after D20) and subsequent increase in strength (from D80) may be due to the

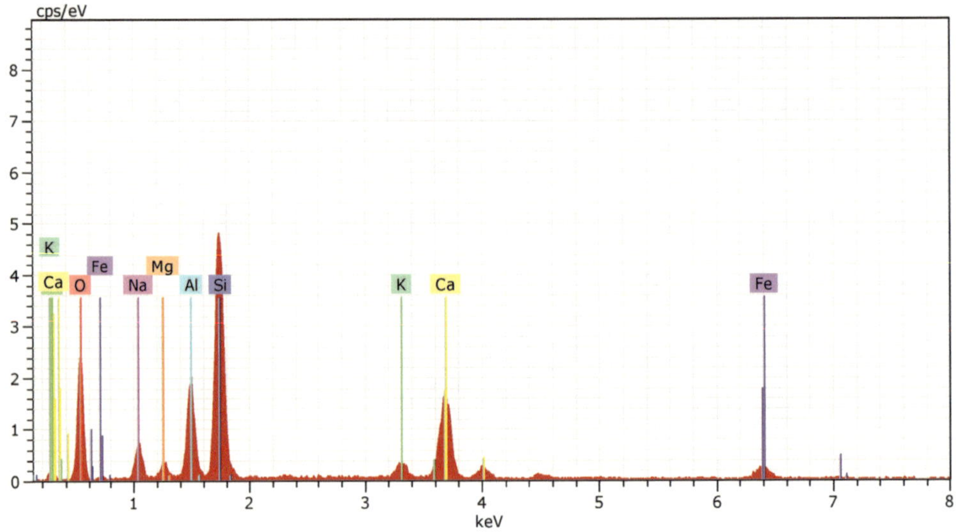

Figure 13.
Elemental composition of G10A.

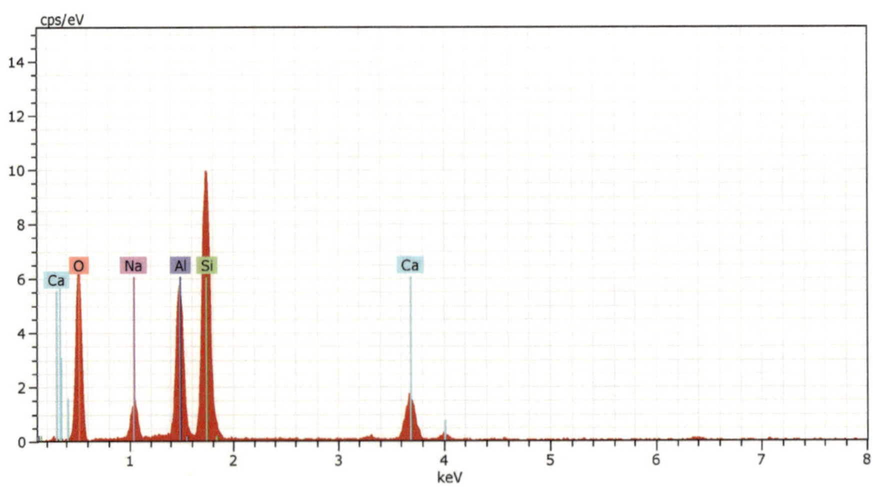

Figure 14.
Elemental composition of G20A.

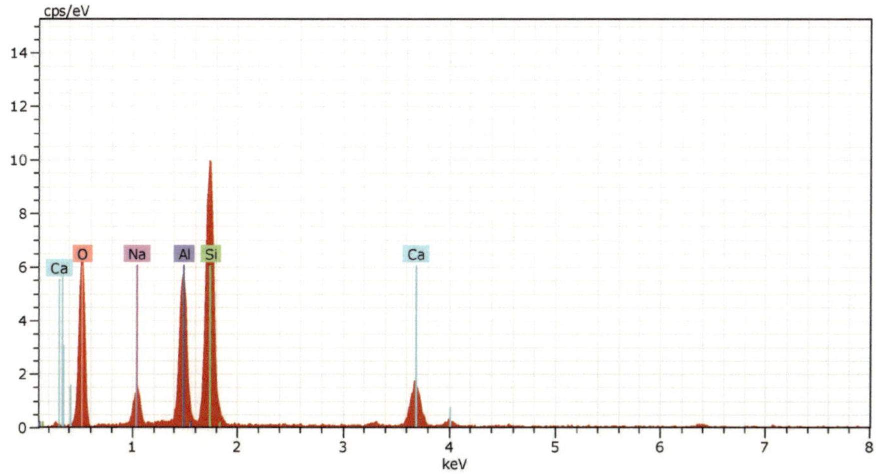

Figure 15.
Elemental composition of G30A.

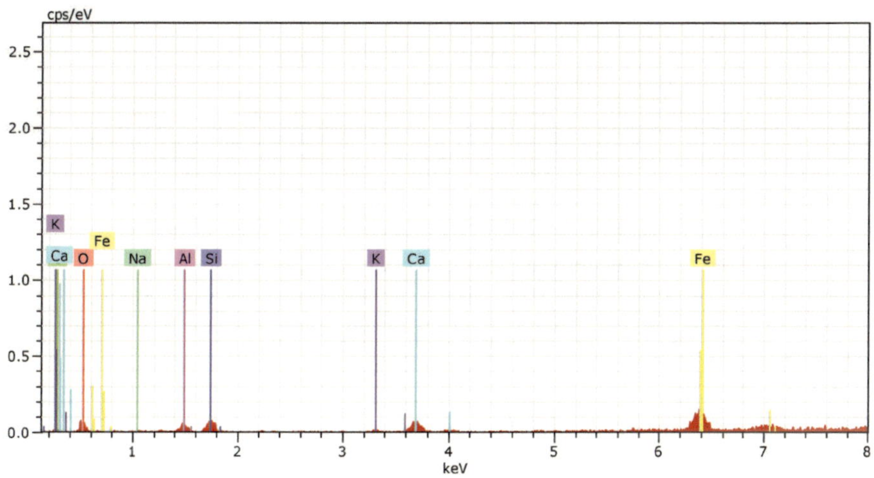

Figure 16.
Elemental composition of G40A.

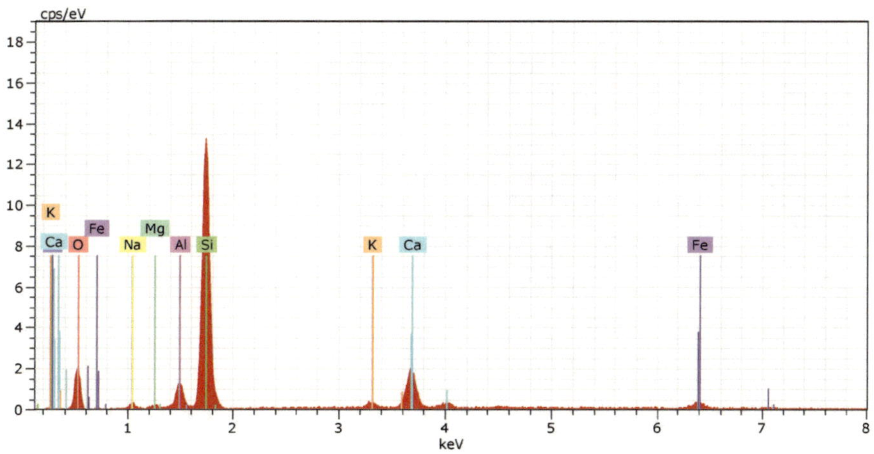

Figure 17.
Elemental composition of G50A.

S. No.	Elemental Composition	G10A	G20A	G30A	G40A	G50A
1.	Si	10.50	15.16	17.58	7.24	27.32
2.	Al	4.32	8.57	7.81	5.56	10.58
3.	Na	4.03	4.98	4.20	1.68	1.73
4.	Ca	3.16	4.31	7.90	4.93	3.94
5.	O	57.49	69.19	69.39	57.19	46.77
Si/Al		2.43	1.768	2.25	1.302	2.58
Na/Al		0.383	0.328	0.537	0.302	0.164

Table 2.
Elemental composition.

above reason. Si/Al ratio confirms the structure of sialate link as (Sialate- Di Silaxo-D20), (Silate Silaxo-D40), (Silate Tri Silaxo-D60), (Silate Silaxo-D80) and (Silate Tri Silaxo-D100). Si/Al ratio of 1 indicates that the structure is Sialate Silaxo, Si/Al ratio of 2 indicates that the structure is Silate Di Silaxo and Si/Al of 3 represents that it is Silate Tri Silaxo.

3.5.2 SEM analysis

SEM analysis was carried out for the samples having mixture of fine aggregates (DMS and M-Sand) and for sample with DMS only. The obtained SEM images are shown in **Figures 18** and **19**.

From **Figure 18**, it shall be observed that geopolymer matrix is well formed and the ITZ gap between the aggregates and matrix is less leading to greater compressive strength in case of D20 which confirms the inference obtained earlier that strength decreases after D20. Further, micro cracks are observed in case of D100. (**Figure 19**).

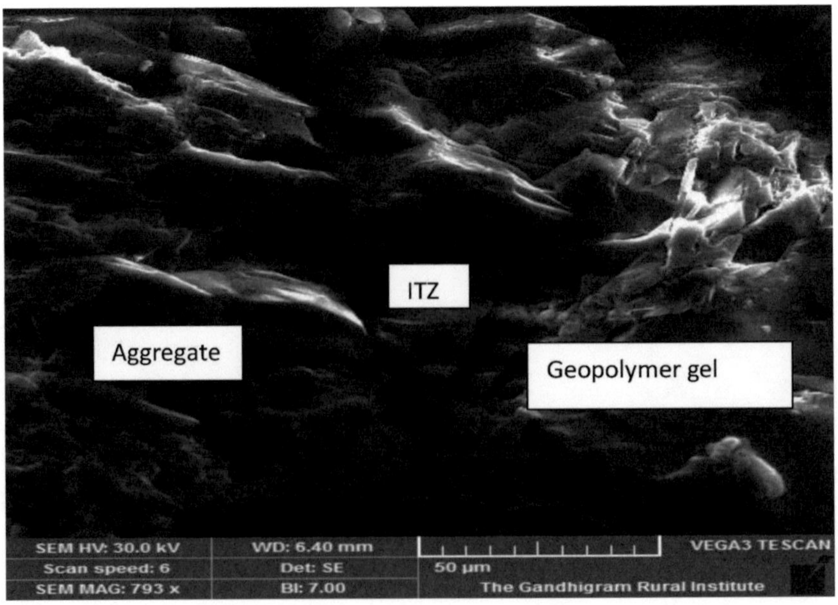

Figure 18.
SEM analysis of G40A.

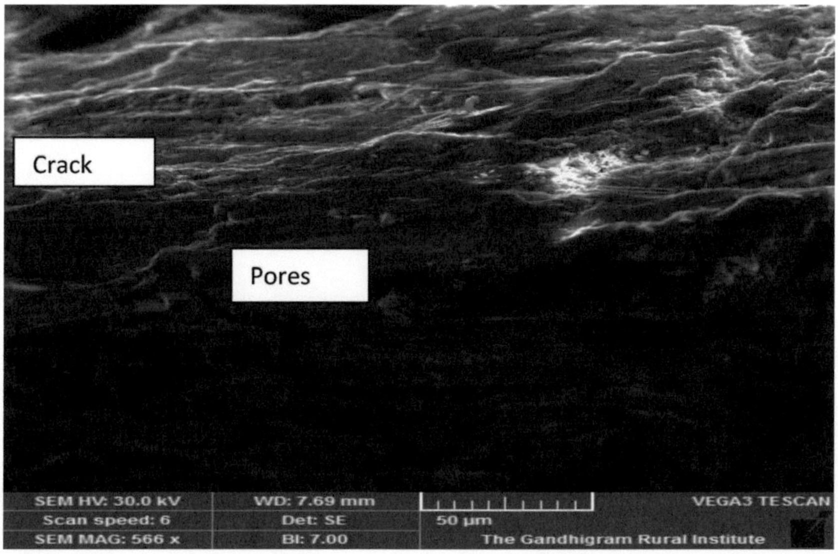

Figure 19.
SEM analysis of GOA.

4. Conclusion

This project mainly focused on arriving at the optimum percentage replacement of flyash by GGBS in geopolymer concrete. The following conclusions are derived from the investigations carried out on compressive strength, split tensile strength, density measurements and flexural strength:

- Compression, split tension as well flexure yielded the same inference that increase in strength was upto 40% replacement of fly ash by GGBS beyond which strength got decreased due to the fact that microstructure gets altered as evidenced by EDAX and SEM analysis.

- Workability was observed to be stiff upto 40% replacement requiring use of superplasticisers at the time of casting.

- Strength values of specimens subjected to alternate wetting drying conditions were slightly lower compared to ambient cured specimens.

Author details

M. Indhumathi Anbarasan[1*], S.R. Sanjaiyan[2] and S. Nagan Soundarapandiyan[3]

1 Department of Civil Engineering, Ramco Institute of Technology, Rajapalayam, Tamilnadu, India

2 Department of Civil Engineering, Tagore Institute of Engineering and Technology, Salem, Tamilnadu, India

3 Department of Civil Engineering, Thiagarajar College of Engineering, Madurai, Tamilnadu, India

*Address all correspondence to: indhumathi@ritrjpm.ac.in

References

[1] T. VenuMadhav, I.V. Ramana Reddy, Vaishali G. Ghorpade and S. Jyothirmai, "Compressive strength study of geopolymer mortar using quarry rock dust" – Construction and Building Materials (2018).

[2] Khoa Tan Nguyen, Tuan Anh Le and Kihak Lee, "Evaluation of the mechanical properties of sea sand-based geopolymer concrete and the corrosion of embedded steelbar" - Construction and Building Materials (2018) 462-472.

[3] Marland, G., T.A. Boden, R.C. Griffin, S.F. Huang, P. Kanciruk and T.R. Nelson (1989),). Estimates of CO2 Emissions from Fossil Fuel Burning and Cement Manufacturing

[4] Dumitru Doru Burduhos Nergis et al. (20200, "XRD and TG-DTA Study of New Alkali Activated Materials Based on Fly Ash with Sand and Glass Powder", Materials, 13(2), 343

[5] M. Albitar, M.S. Mohamed Ali, P. Visintin and M. Drechsler, "Durability evaluation of geopolymer and conventional concretes" - Construction and Building Materials (2017) 374-385.

[6] IS 383: 1970, "Specification for coarse and fine aggregates from naturalsources for concrete".

[7] U.S. Agrawal, S.P. Wanjari, D.N. Naresh, "Impact of replacement of natural river sand with geopolymer fly ash sand on hardened properties of concrete" - Construction and Building Materials (2019) 499-507.

[8] N A Lloyd and B V Rangan "Geopolymer concrete with flyash" Main Prodeedings, Second International conference on Sustainable construction materials and technologies June 28-june 30, 2010.

[9] Muhammad N.S. Hadi, Haiqiu Zhang, Shelley Parkinson, "Optimum mix design of geopolymer pastes and concretes cured in ambient condition based on compressive strength, setting time and workability" - Journal of Building engineering (2019).

[10] K. Jeevanandan and V. Sreevidya, "Experimental investigation on concrete and geopolymer concrete" - Materials Today (2019).

[11] Aissa Bouaissi , Long-yuan Li, Mohd Mustafa Al Bakri Abdullah and Quoc-Bao Bui, "Mechanical properties and microstructure analysis of FA-GGBS-HMNS based geopolymer concrete" - Construction and Building Materials 210 (2019) 198-209.

[12] IS 5816:1999, "Compressive and Splitting tensile strength of concrete".

[13] Mithanthaya I.R. , Shriram Marathe, N B S Rao, VeenaBhat, "Influence of superplasticizer on the properties of geopolymer concrete using industrial wastes" - Materials Today (2017) 9803-9806.

[14] S. JeevaChithambaram, Sanjay Kumar and M.M. Prasad, "Thermo-mechanical characteristics of geopolymer mortar" - Construction and Building Materials 213 (2019) 100-108.

[15] Indhumathi Anbarasan and Nagan Soundarapandian, 2019,"Investigation of mechanical and microstructural properties of geopolymer concrete blended by Dredged Marine Sand and Manufactured Sand under ambient curing conditions", Structural Concrete, DOI 10.1002/SUCO 201900343, Vol.21, No.3, Nov 2019, pp 992-1003

Chapter 5

Incorporation of Phase Change Materials and Application of 3D Printing Technology in the Geopolymer Development

Ahmed Nmiri

Abstract

The building sector accounted for the largest share of both global final energy use and energy-related CO_2 emissions. Despite the efforts made during the last decade to reduce energy consumption and greenhouse gas emissions, the demand for energy is increasing steadily. Thus, development of novel strategies to reduce energy costs and save the environment through a new building regulation has critical importance. Several new technologies are emerging to help achieve the aim of reducing energy usage in building sectors, eliminating greenhouse gas emissions, and recycling waste. Some of these technologies are: (1) the development of geopolymer binder that may be used as an alternative to ordinary Portland cement, (2) the adoption of three-dimensional (3D) printing technology in the civil engineering, and (3) the integration of phase change materials (PCM) in cementitious materials to increase energy efficiency of buildings. In this chapter we review some research about phase change materials-based geopolymer cement, and the adoption of the additive manufacturing technology in geopolymer applications, as well as, point to further areas of study required for wide-scale industry adoption.

Keywords: Aluminosilicate materials, Alkaline activator, Geopolymer, Phase change material, 3D printing

1. Introduction

Due to some advantageous properties of thermal conductivity, high density, and high mechanical strength of Portland cement concrete (PCC), it is a frequently used concrete for applications utilizing PCMs in especially microencapsulated form [1–3]. However, carbon dioxide emission during the production of PCC causes a negative effect on the environment. Compared to PCC, geopolymer concrete (GPC) has several beneficial properties as PCC, but also higher initial strength [4], superior acid resistance [5], high fire resistance [6], and shorter setting time [7, 8]. These features of GPC make it an alternative binder for preparation of PCM containing cementitious materials to be considered for improving building energy efficiency.

Phase change material (PCM), the thermal energy storage (TES), is one of the promising methods used to reduce the environmental impact and to increase energy efficiency of buildings. Phase change materials (PCM) are latent heat storage materials, which can store and release large amounts of energy during a phase change that can occur according to one of many matter transitions (solid–liquid or solid–gas or liquid–gas or solid–solid). However, the most commercially viable transition is between the liquid and solid phases. When the temperature rises above melting point of PCM, this last one melts and absorbs heat, when the temperature drops below melting point; the PCM solidifies and release heat. Heat can also come from other sources such as non-air-conditioned buildings and industrial machinery. There are three main types of PCM: organic, inorganic and eutectic. The most commercially viable PCM is organic since it is chemically stable, safe and nonreactive, does not loss its effectiveness with cycling, can be microencapsulated and has a wide temperature range [9]. The addition of PCM into building materials leads to store the excess of the outdoor environment heat (During the day) and reduce heat transfer to the indoor side of the concrete wall. While, during the cold periods (or at night) the PCM release the stored heat into the building if the inside temperature is too low, and thus causing an increase comfort level in building through providing heating in the winter and providing cooling in the summer without the use of an air-conditioning system. Thereby reduce the fossil fuels-based energy consumption.

Most of the investigations were focused on the addition of microencapsulated PCM (MPCM) to the standard concrete recipes. The literature survey indicated that the combination of MPCM with cementitious material resulted in low composition fraction due to the fact that more loading is resulted in the final product with low TES capacity and low mechanical strength. Another problem is the leakage of PCM caused by breaking some parts of capsules during mixing and compression processes. Moreover, the combination of MPCM with GP is not cost effective because of the high cost and complex synthesis nature of MPCM.

Moreover, microencapsulated or encapsulated phase change materials are particles consisting of a core material the (PCM) and an outer wall (shell). The shell is an inert, stable polymer or plastic or metallic [10]. It does not melt under normal processing and use conditions. The shell acts as a barrier between the core material and the surrounding matrix and controls the volume change of the PCM during its phase change. Due to the very high cost of PCM metal-based encapsulations other method has been adopted to avoid the high cost of the encapsulation of PCM. This method consists in direct incorporation of non-encapsulated PCM into concrete materials. However, this method leads to the leakage of PCM during its liquid state, as well as the corrosion of the concrete matrix due to the corrosive nature of some PCMs. Not to mention the corrosion of the exposed surfaces of the concrete matrix and the reinforcement bar embedded in concrete causing by the environment (air, humidity, salt water, or other hostile environment), which enhances the porosity and permeability of the materials, thereby reducing its mechanical and structural properties.

Because of its cost-effective and eco-friendly, 3D printing is an excellent method to create a building with an efficient thermal regulation and make the integration of PCM in building more efficient, less expensive and more environment friendly [11]. However, among the obstacles for application of 3D printing in geopolymer-based construction are the low mechanical strength, long setting time, and low tensile strength [12, 13].

2. Geopolymers

Geopolymer cement (GP) has several beneficial properties as Portland cement (PC), but also it is much more ecological. It is known as green material [14], because of its low energy consumption, and low emission of polluting gases during manufacture, which makes it an alternative binder for cementitious materials. Geopolymer is produced from the activation of aluminosilicate materials such as metakaolin [15], blast furnace slag [16], fly ash [17] and so on, by a homogenous solution of alkali hydroxide and alkaline silicate, at ambient conditions [14, 18]. The mechanism consists of (1) the dissolution of the aluminosilicate structure under the effect of the alkali hydroxide to form oligomers of silicate $[SiO(OH)_3]^-$ and aluminate $[Al(OH)_4]^-$, and (2) the condensation of these free oligomers under the effect of the alkaline silicate to form another amorphous three-dimensional network [14, 18, 19]. The role of Na^+ and K^+ ions consists in balancing the electrical charge of Al^{3+} in IV-fold coordination [20, 21]. Geopolymer materials can be used for encapsulating heavy metals. In fact, the charge-balancing-alkali ions in the geopolymer network can be partially replaced, by ion exchange, with radioactive elements, and reducing them migration into the environment.

Alkali silicate solutions, also called water glass, such as sodium silicate, potassium silicate, lithium silicate etc. are consolidating agents for the material, they increase the formation rate of tetrahedrally coordinated aluminum [22] and improve the mechanical and physical properties of the final materials. Alkali silicate solutions have many applications, such as the production of silica gel and formulation of refractory ceramics and cements. The use of a highly reactive alkaline silicate solution and a highly amorphous aluminosilicate material enhances the formation of the geopolymer network [22]. The reactivity of alkaline silicate solutions was found to be dependent on the SiO_2/Al_2O_3 and SiO_2/Na_2O ratios. According to some literatures [15, 23–25], the highest mechanical properties of geopolymers is obtained when SiO_2/Al_2O_3 molar ratio is between 3.0 and 3.8, Na_2O/Al_2O_3 molar ratio is about 1 and the NaOH solution concentration is approximately 10 M. The SiO_2/Al_2O_3 ratio may change with the type of raw material used as source of alumino-silicate [17] and then impacts on geopolymers properties [26]. The physical properties of geopolymers are improved when SiO_2 are added to the mixture [14, 27]. Kong and Sanjayan [28] stated that alkaline solution selection and SiO_2/Al_2O_3 ratio are critical parameters necessary to optimize the performance of geopolymer (at ambient or at elevated temperature).

Recent studies show that use waste can offer an alternative to alkaline silicate with potential advantage of lower cost and lower environmental impact. Torres-Carrasco and Puertas [21] have studied the feasibility of using industrial waste glass as a source of silica to replace sodium silicate in the alkaline activation of fly ash. They found, according to the analysis of the mechanical properties, degree of reaction and microstructure of alkali-activated fly ash, that the dissolved waste glass silicate by the NaOH solution had a substantial impact on the composition of the geopolymerisation reaction. Other studies demonstrated that usage of waste glass in concrete enhances its acid resistance as well as its physical and mechanical properties [29, 30]. In contrast the addition of waste glass reduces the plasticity of the fresh paste, thereby reduce its workability, thus a super-plasticizer is needed [31].

Tong et al. [32] investigated the production of sodium silicate solution from Rice Husk Ash (RHA). A hydrothermal process for the dissolution of RHA in sodium hydroxide solution was developed. Optimized procedure parameters were found to be: NaOH concentration 3 M, heating temperature 80°C and heating duration 3 h. The obtained solution was used for the production of alkali-activated binder made with a blend of fly ash and ground granulated blast furnace slag. Obtained compressive strength of mortar was in the range of 60 MPa at 28 days, which confirmed the equivalence between the solution produced with the optimized method and commercially available options. Cost analysis indicated that the proposed method could allow a reduction of almost 55% of the cost for the activation of alkali-activated binder.

Silica fume has been used as activator in metakaolin-geopolymer preparation [24], the Mk-based geopolymer with a silica fume content of 6 wt% (compared with those with 2% and 10%), corresponding to a SiO_2/Al_2O_3 molar ratio of 3.84, resulted in the highest compressive strength, which was explained based on its high compactness with the smallest porosity. Silica fume improved the compressive strength by filling interstitial voids of the microstructure because of its fine particle size.

Geopolymer materials are differentiated by their SiO_2/Al_2O_3 ratio that affects their structure and application. Aluminosilicate-based GPs are designated by the term "poly (sialate)," which is an abbreviation of poly (silico-oxo-aluminate) or $(\text{-Si-O-Al-O})_n$. The various types of poly (sialate), according to Davidovits [14], are polysialate or (-Si-O-Al-O-) or (PS) (with $SiO_2/Al_2O_3 = 2$), poly (sialate-siloxo) or (-Si-O-Al-O-Si-O-) or (PSS) (with $SiO_2/Al_2O_3 = 4$) and poly (sialate-disiloxo) (-Si-O-Al-O-Si-O-Si-O-) or (PSDS) (with $SiO_2/Al_2O_3 = 6$). Each type of GP cement possesses special features: good thermal insulation for PS, high strength and good solidification in presence of toxic waste for PSS, excellent fire resistance and high adhesion for PSDS [14, 33].

Silva et al. [15] revealed that the properties of GP systems can be drastically affected by minor changes in the available SiO_2 and Al_2O_3 concentrations during synthesis. The best resistance of metakaolin-based GPs was obtained when the SiO_2/Al_2O_3 molar ratio is between 3.0 and 3.8. These ratios could however change according to the type of raw material used as a source of aluminosilicate. Yunsheng et al. [34] found for example the highest compressive strength (34.9 MPa) for metakaolin-based GPs with a higher SiO_2/Al_2O_3 ratio (equal to 5.5).

Burciaga-Diaz et al. [35] investigated the compressive strength evolution as a function of the curing temperature at 20 and 80°C, and using the molar ratios of SiO_2/Al_2O_3 (2.64–4.04) and Na_2O/Al_2O_3 (0.62–1.54). The results revealed that the optimal ratios that yielded the greatest compressive strength were $SiO_2/Al_2O_3 = 2.96$ and $Na_2O/Al_2O_3 = 0.62$, 0.93. For greater SiO_2/Al_2O_3 ratios, good final compressive strengths were registered, but the setting time was very long. Curing at 80°C for 24 h was favorable for a rapid strength gain only at early ages, however, at later ages, the highest compressive strengths were obtained after curing at 20°C (this temperature makes it possible to avoid any thermal degradation of the geopolymer). Catauro et al. [36] investigated the structure and the mechanical behavior of the organic–inorganic hybrid materials consisting of an inorganic matrix of a metakaolin-based GP in which the polyethylene glycol (PEG) was added as a plasticizer. The results revealed that the elastic strain increased for a fixed value of stress with the percentage of PEG increases but a decrease of flexural and compressive strengths due to the increasing of porosity. PEG-free samples can reach final mechanical resistance faster than hybrid systems.

3. PCM and construction materials

Numerous studies have investigated the integration of PCM to improve the building energy efficiency. Shadnia et al. [37] studied the mechanical and thermal properties of fly ash-based GP mortar containing different amount of PCM in the form of micro-encapsulated powder. They found that the unit weight and compressive strength of the GP mortar decreased when more PCM was incorporated due to lightweight, small shear strength and stiffness of PCM. In the meantime, the effect of the melting of PCM on the strength of GP mortar is negligible. Cao et al. [7] investigated the effect of microencapsulated PCM (paraffin Rubitherm) on the thermal performance and compressive strength of PC concrete and GP concrete. The results revealed that the increasing of the amount of micro-encapsulated PCM (MPCM) improved the latent heat, lowered the thermal conductivity, which could improve the thermal performance of building materials, and decreased the compressive strength in GP concrete compared to PC concrete due to the enhancement of porosity. The results also revealed that GP concrete exhibited better energy saving properties than PC concrete at the same conditions.

In another investigation, Duy Cao et al. [38] conducted a more detailed study on the effect of the polymer shell on the microstructure, thermal, and mechanical properties of GP concrete by using different kinds of MPCMs. They have found that the integration of PCM with a polymer shell containing polar functional groups into GP concrete at 5.2% by weight shows the best thermal performance but the lowest compressive strength due to the largest increase of GP concrete porosity and better interface bonds between microcapsules and the concrete matrix.

In order to know and examine the cause of low compressive strength after adding PCM, Pilehvar et al. [39] investigated the effect of incorporation of MPCM in solid and liquid states on the mechanical properties and microstructure of both GP and PC concretes. The results revealed that the compressive strength of both GP concrete and PC concrete decreased with increasing amount of MPCM. This decrease might be caused by the lower stiffness and strength of MPCM compared to sand, causing MPCM to be deformed or broken during the compression test [37]. It is also possible that air gaps, low adhesion and weak bonds between MPCM and the surrounding matrix may contribute to the strength reduction. The results also revealed that whether the PCM was in solid (Temperature below melting point) or liquid (temperature above melting point) state did not significantly affect the mechanical properties of GPC, while melting the PCM was found to reduce the strength of PC concrete. This was due to the fact that GP concrete has higher compression strength than PC concrete.

GP has been considered as an alternative carrier matrix to hold PCM in macro or micro encapsulated form as well as their being of more environmentally friendly material. Jacob et al. [33] fabricated encapsulated phase change materials EPCM (10 mm) consisting of molten chloride salt as a core encapsulated by fly ash geopolymer-based shells. They employed tow methods (dip-coating and pre-formed shells) to fabricate the geopolymer capsules. They found that the dip-coated capsules resulted in non-uniform shell thickness and shape, which promotes uneven heat transfer and high stress areas during phase change. However, the preformed capsules method allowed a greater control over size, shape and shell thickness.

Frattini et al. [40] synthesized GP mortar containing an expanded clay aggregate (supporting-shape) incorporated paraffine PCM. The results showed that with the paraffin content up to about 30–40% by weight, the matrix exhibited good mechanical properties and very high fire resistance.

Sukontasukkul et al. [41] studied the mechanical properties and heat insulation of wall panels made of fly ash-GPC containing porous lightweight aerated block (9.5–19 mm) impregnated with paraffin PCM. The results showed that the incorporation of PCM aggregate improved both thermal storage and heat insulation of GP panel. The results also showed that the density of GP increased with the increase in paraffin content in aggregate. Nonetheless, since lightweight aggregate was much weaker than GP paste, the increase in lightweight aggregate content caused the strength reduction.

Wang et al. [42] studied the influence of PCM on mechanical and thermal properties of clay GP mortar, PCM has been prepared by vacuum method using paraffin as the heat-absorbing material, expanded perlite as the supportive material, and two encapsulation methods were developed by using $CaSiO_3$ and Na_2SiO_3 as the capsules, and the corresponding PCMs were incorporated with clay GP mortar. They found that the incorporated PCM in clay GP mortar effectively reduced the transport heat. In addition to this, expanded perlite plays good package effect on paraffin during the process of solid–liquid phase change, which can guarantee the composite PCM in solid conditions during the solid–liquid phase change in the macro. By using $CaSiO_3$ and Na_2SiO_3 as the capsules, the PCM building materials can effectively avoid the leakage of paraffin.

In order to obtain effective composite PCMs for thermal energy storage purpose in buildings, Sarı et al. [3] developed a form-stable composite PCM composed of the cement impregnated with the eutectic mixture of capric acid(CA)-myristic acid(MA) as PCM through vacuum impregnation method. And then they investigated its chemical structure. The FTIR spectra/XRD patterns not showed any new band/peak. This result confirms the fact that any chemical reaction is not carried out between the cement matrix and form-stable composite PCMs.

Kastiukas et al. [43] performed an investigation to determine the most efficient coating method and material regarding its ability at retaining the PCM. They produced macro-encapsulated aggregates using expanded clay lightweight aggregates (in the 2–10 mm size range + numerous small and large pores) as supportive shape impregnated with paraffin PCM. The different coating materials used were: Sika Latex, Weber dry-lastic and polyester resin adhesive (Palatal). The macro-encapsulated aggregates were then used as aggregates in GP binders made from a combination of aluminosilicate rich mud and waste glass. The results revealed that PCM vacuum impregnation was very successful for expanded clay lightweight aggregate. The polyester resin was determined to be the most suitable choice of coating material for the PCM impregnated lightweight aggregates. Polyester resin coating was also chemically stable and neutral, and improving thermal conductivity.

As conclusion several methods have been proposed to incorporate PCM in buildings:

1. The first method is direct incorporation of PCM into concrete materials at the time of mixing [44]. In this case, PCM is distributed freely in concrete. However, this method leads to the leakage of PCM during their liquefaction. This causes contamination of the host material, PCM loss and a reduction of the thermal energy storage capacity and mechanical properties of the building materials.

2. The second method consists in microencapsulation of PCM into small closed sphere capsules called microencapsulated phase change materials MPCM to mixing them later with the fresh concrete [7]. This method prevents leakage. MPCM improve thermal energy storage. However, this method is very expensive and the deterioration of MPCM during the mixing process may cause leakage of PCM into the building materials. In addition, a poor compatibility between MPCM and the concrete matrix, and the tendency of MPCM to agglomerate may be the main causes of increased porosity, and therefore a decrease in compressive strength of the building materials [45].

3. The third method consists in direct immersion of the cementitious powder in liquid PCM until obtain a form-stable PCM material [46]. This method is simple and low cost and allows for better dispersion of PCM. However, the stress of the volume expansion during phase change process leads to enhance

the porosity in buildings, and crack the concrete matrix, thereby causing the concrete matrix to crack and reduce mechanical strength.

4. Another method is the impregnation of PCMs in porous supportive shape materials through vacuum impregnation to obtain a macro-encapsulated PCMs and then mixing them with fresh concrete [33]. This method controls the volume expansion during phase change process. However, the solidification of PCM only around the edges during the time of heat regaining from liquid state prevents effective heat transfer. Overall, encapsulated PCM may be spherical, rectangular, cylindrical or irregular. It was found that spherical capsules resulted in the highest heat transfer rate. Also, the small size of encapsulated PCM leads to an increase in the heat transfer rate [33]. Moreover, the small encapsulated PCM can fill the cavities between aggregates and sand, thereby reducing the concrete porosity [7].

Therefore, there is strongly needed to develop geopolymer material with thermal energy storage TES ability besides without including these difficulties. In order to make this product more efficient, cheaper, and more environmentally friendly, the 3D printing technology could be used [11, 13].

4. Additive manufacturing

Additive manufacturing technology, also called 3D printing technology, is also one of the promising methods used to reduce the energy demand and to reduce greenhouse gas emissions, in fact the application of this method takes less time than conventional manufacturing, and it uses the precise amount of material needed for the construction, which reduces the material costs, waste and negative impact on the environment [47, 48]. The digitally controlled construction process is done by a 3D printer. It consists in placing the fresh paste inside the loading container. Then, this last one is installed on the construction robot to start the printing process. The movable extrusion head is controlled by software to move in x, y, and z directions. The printer nozzle, on the extrusion head, traces the desired shape, layer by layer, until the final structure is achieved. The printing conditions could be optimized by modifying the nozzle diameter, the extrusion speed, layer heights, and the time gap between each layer [47–49].

The development of building materials adapted to the construction-based 3D printing technology has become of great importance in the world since the beginning of the second decade of the 21st century [11, 50].

3D printing process depends on two main factors, namely flowability and buildability [51]. The flowability is the fresh concrete extrusion capacity through the nozzle of the printer. The buildability is the ability of fresh concrete to retain the desired shape and hold layer overlays without collapsing. The buildability depends on the setting time and the mechanical property of the fresh material. Flowability and buildability can be considered as against each other, indeed if flow is increased, buildability decreases, and vice versa. Which is the main obstacle to applying the 3D printing method in the building construction [11, 52].

The mechanical activation of raw materials is one of many solutions used to overcome these obstacles. The mechanical activation reduces the setting time and improves the mechanical performance of the resulting material [53]. For example, the increasing of grinding time of the raw materials breaks the crystalline structure of the inert material and increases its reactivity which improves the mechanical

properties of the hardened product. However, overgrinding leads to a decrease in fresh material flowability, which causes further demand of water to reincrease the flowability [54]. Then the evaporation of water during curing process increases shrinkage and cracks in the resulting material which causes a decrease in its mechanical properties. If water does not added, the interaction between activators and aluminosilicate material remains low, thereby the unreacted materials causes cracks and affects the mechanical properties of the resulting material [55].
To increase the flowability and workability of fresh past while maintaining its mechanical performance, plasticizers should be used [56].

To reinforce the geopolymer material, Ma et al. [57] entrained a continuous micro steel cable during filaments deposition process, which demonstrated significant improvement of mechanical strength, toughness and post-cracking deformation of geopolymer composite. Shakor et al. [58] added lithium carbonate to reduce the setting time for the cement mixture.

5. Wall coating material with anti-corrosion and anti-leakage properties

Wall coatings are decorative or protective layer that are applied to the interior or exterior surfaces of walls. Coating is applied to the surface of the wall via different methods depending on the nature of that wall and the nature of the coating itself. Among these methods are plastering [59], paint brush [60], roller [61], air spray [62] and etc. Coating designed for thermal energy storage and thermoregulating should be corrosion resistance, anti-leakage and anti-heat [63–65].

Corrosion is the gradual destruction of the concrete matrix by chemical reaction with the environment [66, 21]. This reaction enhances the porosity and permeability in the concrete matrix. Moisture, CO_2, chloride, and other harmful ions could reach the surface of reinforcement bars, thereby causing corrosion of the bars and reducing the mechanical and structural properties of the concrete.

To reduce the water absorption and chloride diffusion coefficient, Zheng et al. [67] coated the concrete with Epoxy resin nanocomposites containing 0.3 wt% of graphene oxide. The chloride ion penetration resistance is due to the formation of crosslinking in the composite coating, improvement of hydrophobicity and shielding effects of graphene oxide.

To prevent penetration of hostile elements, and early crack of concrete, epoxy resin nanocomposites modified with graphene oxides (GOs) were prepared using a solution blending process and then sprayed onto testing blocks of concrete [68].

Waterborne silicate coating is an anticorrosive coating used to protect steel bar in the concrete, it consists of alkali metal silicate, rust-proof pigment, and modification material (to modify the alkali metal silicate solutions) such as acid modification, silicone acrylic emulsion, styrene acrylic emulsion... [69, 70]. The waterborne silicate coating has excellent corrosion resistance, good acid-alkali resistance and high heat resistance [71]. The zinc silicate works with the alkali metal silicate, forming a dense and stable film on the metal surface, and reducing the penetration rate of water and other ions. Coatings for reinforcement bars are widely available. In addition, the reinforcement bars are embedded in the concrete matrix and they do not expose to the hostile environment. Moreover, geopolymer materials have high corrosion resistance. The coating of the reinforcement bars may be not necessary The interaction between geopolymer-based coating material and the superficial elements of reinforcement bar or mild steel leads to formation of passive layer which prevent reinforcement bar and mild steel from corrosion [66]. Fly ash-slag geopolymer has good corrosion resistance and low corrosion rate compared to fly ash geopolymer [72].

Afshar et al. [73] reported that the zinc-rich epoxy primer as a coating on mild steel rebar has the best performance when used in combination with concrete containing 25% fly ash, 10% silica fume and 3% inhibitor by cementitious material weight.

Zhao et al. [74] believed that adding ultrafine silica powders dispersed by hexamethyldisilazane (HMDS) and polyacrylic acid PA emulsion improved the water, acid, alkali, heat and aging resistance of the polymer modified cementitious coatings PCCs. The appropriate amount of the modified ultrafine silica powders is about 5% of the mass of the PA emulsion, because higher water absorption and decreased tensile properties happened when the amount is too large (10%). Hexamethyldisilazane (HMDS) disperses the ultrafine silica SiO_2 powders and reduces their agglomeration. SiO_2 chemically interacts with the polyacrylic acid PA emulsion to form a cross-linked network structure. It was found the PCCs with 4 wt% HMDS modified SiO_2 powders had fewer micro-defects and more compactness, thus the tensile properties and durability under different conditions, such as water, acid, alkali, heat and artificial aging, were significantly improved. The water absorption and chloride ion permeability coefficient of concrete coated by the PCCs with 4 wt% well-dispersed SiO_2 powders were also decreased.

Xu et al. [75] developed a colorful and robust superhydrophobic concrete (CSC) coating composed of cement, sand, water-based stone protector, and dyes, and meets both performance and esthetic requirements. And they reported that this coating exhibits excellent chemical durability and weather resistance, and has promising application prospects on the outside wall of concrete.

Several studies demonstrated that usage of glass waste in concrete enhances its acid resistance as well as its physical and mechanical properties. Bisht et al. [29] have used waste of glassy materials made of soda lime to produce acid resistance concrete. They have shown that the best optimized performance in terms of compressive strength and acid resistance can be obtained by substituting up to 21% of sand by glass waste. The overadding of glass waste above to 21%, although it increases the acid resistance, it enhances the porosity which decreases mechanical strength of the concrete. Glass waste can also be used as aggregate or as source of silicate in the manufacturing of geopolymer concrete [30]. In contrast the addition of glass waste reduces the plasticity of the fresh paste, thereby reduce its workability, thus a super-plasticizer is needed [31].

Morefield el al. [76] have developed a novel coating that is based on hydraulically reactive silicate cement blended with a glass enameling frit and fused onto the steel reinforcement: If the enamel coating is cracked the freshly exposed calcium silicate cement grains will react with any humidity in contact with them to produces a cement paste in the crack.

6. Discussion and conclusion

As can be seen from the literature, the PCMs were incorporated by building materials in two ways: (i) addition of MPCM to building materials (ii) the addition of non-encapsulated PCM to building materials by impregnation (directly or by vacuum) method. Despite numerous studies carried out to make fresh materials suitable for 3D printing, no study has developed geopolymer-based PCM for 3D printing construction.

- Although the vermiculite and perlite clay minerals have been used to prepare FSCPCMs with leakproof property until now [77, 78], vermiculite and perlite based FSCPCMs has not been yet integrated with fly ash based-GP to prepare novel kinds of GP-FSCPCM concretes which can be used to decrease temperature fluctuations of building inside.

- Literature survey indicated that paraffin as PCMs were commonly used with geopolymers. However, bio-based fatty acid eutectic mixtures are rarely used. When compared to the paraffins, the fatty acid has better TES characteristics in terms of especially subcooling degree, volume change, latent heat energy storage capacity, phase change reversibility and low-cost due to their produce ability from the vegetable and animal fats [79, 80].

Therefore, based on these considerations, we think that the most proper way to achieve GP-FSCPCM concrete with the most effective is incorporation of non-encapsulated PCM with proper material by vacuum infiltration method and then addition to the GP mortar.

To prepare form-stable composite phase change materials (FSCPCMs), PCMs may be impregnated separately with vermiculite and perlite using vacuum impregnation technique (**Figure 1**) [78, 81].

To achieve the form-stable composition, the mass fraction of PCM could be changed. Then, each of the vermiculite/PCM and perlite/PCM composite samples could be subjected to leakage test by heating them above melting temperature of regarded PCM. After this test, the composite with free of leakage will be defined as FSCPCM.

Different clays such as diatomite, perlite, kaolinite, bentonite, vermiculite etc. as porous, lightweight supporting materials have been used to produce form-stable composite PCMs (FSCPCMs) [82, 83]. Among these clay minerals, vermiculite (VMT) and perlite (PER) are good supporting materials for absorbing organic PCMs. VMT is a lightweight material with porous, inexpensive, ecologically harmless, non-toxic and expandable as much as 8–30 times its original size, when heated to about 800°C. Therefore, VMT is used for construction and insulation in buildings. Perlite (PER) is glassy volcanic rhyolitic rock. PER can be expanded up to 10–20 times its original volume when heated rapidly at 850–1150°C. The resulting expanded perlite (EPER) particles are spherical in shape, usually fluffy, highly porous due to a foam-like cellular internal structure. EPER has low sound transmission, high fire resistance, a large surface area, low moisture retention and a very low density. Besides it is classified as environmentally safe ultra-lightweight

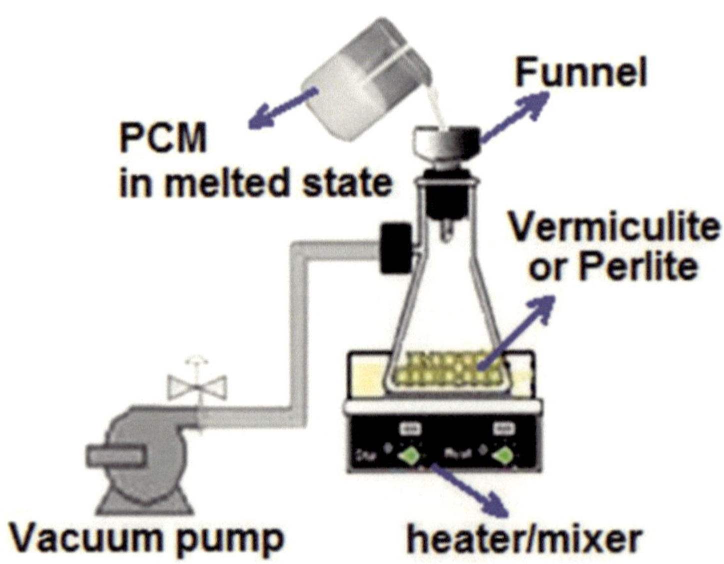

Figure 1.
Vacuum impregnation technique used for preparation of FSCPCMs.

building material. In the buildings, VMT and PER are used as lightweight aggregate for plaster, concrete compounds, firestop mortar, and component of interior fill for walls. Moreover, they have good chemical compatibility with organic PCMs such as fatty acids and their binary mixtures. Therefore, both clay minerals are promising candidates as building material to prepare FSCPCMs for TES applications in buildings [78, 81].

By using, 3D printing technology, GP, PCMs and charging storage facilities with energy generated from renewable sources, we can reduce the greenhouse gas emissions and the dependence on fossil fuels, preserve the environment, attenuate the overheating or excessive cooling of the room and maintain a desirable temperate without the use of the air-conditioning system, allow to positively influencing indoor room temperature by storing direct solar radiation.

Author details

Ahmed Nmiri
Department of Chemistry, Faculty of Sciences of Tunis, University of Tunis El-Manar, Tunis, Tunisia

*Address all correspondence to: nmiriahmed@gmail.com

References

[1] Tyagi V V., Kaushik SC, Tyagi SK, et al. Development of phase change materials based microencapsulated technology for buildings: A review. Renew Sustain Energy Rev 2011; 15: 1373-1391.

[2] Rao VV, Parameshwaran R, Ram VV. PCM-mortar based construction materials for energy efficient buildings: A review on research trends. Energy Build 2018; 158: 95-122.

[3] Sarı A, Bicer A, Karaipekli A, et al. Preparation, characterization and thermal regulation performance of cement based-composite phase change material. Sol Energy Mater Sol Cells 2018; 174: 523-529.

[4] Rajini B, Rao AVN. Mechanical Properties of Geopolymer Concrete with Fly Ash and GGBS as Source Materials. Int J Innov Res Sci Eng Technol 2014; 03: 15944-15953.

[5] Valencia-saavedra WG, Mejía R, Gutiérrez D, et al. Performance of FA-based geopolymer concretes exposed to acetic and sulfuric acids. Constr Build Mater 2020; 257: 119503.

[6] Dupuy C, Gharzouni A, Texier-Mandoki N, et al. Thermal resistance of argillite-based alkali-activated materials. Part 1: Effect of calcination processes and alkali cation. Mater Chem Phys 2018; 217: 323-333.

[7] Cao VD, Pilehvar S, Salas-Bringas C, et al. Microencapsulated phase change materials for enhancing the thermal performance of Portland cement concrete and geopolymer concrete for passive building applications. Energy Convers Manag 2017; 133: 56-66.

[8] Zhang P, Zheng Y, Wang K, et al. A review on properties of fresh and hardened geopolymer mortar. Compos Part B Eng 2018; 152: 79-95.

[9] Abhat A. Low temperature latent heat thermal energy storage: Heat storage materials. Sol Energy 1983; 30: 313-332.

[10] Pitié F, Zhao CY, Cáceres G. Thermo-mechanical analysis of ceramic encapsulated phase-change-material (PCM) particles. Energy Environ Sci 2011; 4: 2117-2124.

[11] Panda B, Paul SC, Hui LJ, et al. Additive manufacturing of geopolymer for sustainable built environment. J Clean Prod 2017; 167: 281-288.

[12] Chan SSL, Pennings RM, Edwards L, et al. 3D printing of clay for decorative architectural applications: Effect of solids volume fraction on rheology and printability. Addit Manuf 2020; 101335.

[13] Ngo TD, Kashani A, Imbalzano G, et al. Additive manufacturing (3D printing): A review of materials, methods, applications and challenges. Compos Part B Eng 2018; 143: 172-196.

[14] Davidovits J. *Properties of geopolymer cements*. Saint-Quentin, France: Geopolymer Institute, 1994.

[15] Silva P De, Sagoe-Crenstil K, Sirivivatnanon V. Kinetics of geopolymerization: Role of Al2O3 and SiO2. Cem Concr Res 2007; 37: 512-518.

[16] Huang X, Yu L, Li DW, et al. Preparation and properties of geopolymer from blast furnace slag. *Mater Res Innov* 2015; 19: S10-413-S10-419.

[17] Van Jaarsveld JGS, Van Deventer JSJ, Lukey GC. The characterisation of source materials in fly ash-based geopolymers. Mater Lett 2003; 57: 1272-1280.

[18] Weng L, Sagoe-Crentsil K. Dissolution processes, hydrolysis and

condensation reactions during geopolymer synthesis: Part I-Low Si/Al ratio systems. J Mater Sci 2007; 42: 2997-3006.

[19] A. Nmiri, N. Hamdi, O. Yazoghli-Marzouk, et al. Synthesis and characterization of kaolinite-based geopolymer: Alkaline activation effect on calcined kaolinitic clay at different temperatures. J Mater Environ Sci 2017; 8: 276-290.

[20] Subaer. *The influence of aggregate on the microstructure of geopolymer, PhD thesis of Curtin University of Technology, Department of Applied Physics*. 2004.

[21] Torres-Carrasco M, Puertas F. Waste glass in the geopolymer preparation. Mechanical and microstructural characterisation. *J Clean Prod*. Epub ahead of print 2015. DOI: 10.1016/j.jclepro.2014.11.074.

[22] Vidal L, Gharzouni A, Rossignol S. Alkaline Silicate Solutions: An Overview of Their Structure, Reactivity, and Applications.

[23] Bernal SA, Rodríguez ED, Mejía de Gutiérrez R, et al. Mechanical and thermal characterisation of geopolymers based on silicate-activated metakaolin/slag blends. J Mater Sci 2011; 46: 5477-5486.

[24] Ahmed Nmiri, Myriam Duc, Noureddine Hamdi, Oumaya Yazoghli-Marzouk ES. Replacement of alkali silicate solution with silica fume in metakaolin-based geopolymers. Int J Miner Metall Mater 2019; 26: 555-564.

[25] Heah CY, Kamarudin H, Mustafa Al Bakri a. M, et al. Kaolin-based geopolymers with various NaOH concentrations. Int J Miner Metall Mater 2013; 20: 313-322.

[26] Komnitsas K, Zaharaki D. Geopolymerisation: A review and prospects for the minerals industry. Miner Eng 2007; 20: 1261-1277.

[27] De Silva P, Sagoe-Crentsil K. The Effect of Al2O3 and SiO2 On Setting and Hardening of Na2O-Al2O3 -SiO2 - H2O Geopolymer Systems. J Austalia Ceram Soc 2008; 1: 39-46.

[28] Kong DLY, Sanjayan JG. Damage behavior of geopolymer composites exposed to elevated temperatures. Cem Concr Compos 2008; 30: 986-991.

[29] Bisht K, Ahmed KIS, Ramana P V. Gainful utilization of waste glass for production of sulphuric acid resistance concrete. Constr Build Mater 2020; 235: 117486.

[30] Avancini TG, Souza MT, de Oliveira APN, et al. Magnetic properties of magnetite-based nano-glass-ceramics obtained from a Fe-rich scale and borosilicate glass wastes. Ceram Int 2019; 45: 4360-4367.

[31] Jimenez-Millan J, Abad I, Jimenez-Espinosa R, et al. Assessment of solar panel waste glass in the manufacture of sepiolite based clay bricks. Mater Lett 2018; 218: 346-348.

[32] Tong KT, Vinai R, Soutsos MN. Use of Vietnamese rice husk ash for the production of sodium silicate as the activator for alkali-activated binders. *J Clean Prod*. Epub ahead of print 2018. DOI: 10.1016/j.jclepro.2018.08.025.

[33] Jacob R, Trout N, Raud R, et al. Geopolymer encapsulation of a chloride salt phase change material for high temperature thermal energy storage. *AIP Conf Proc*; 1734. Epub ahead of print 2016. DOI: 10.1063/1.4949119.

[34] Yunsheng Z, Wei S, Zongjin L. Composition design and microstructural characterization of calcined kaolin-based geopolymer cement. Appl Clay Sci 2010; 47: 271-275.

[35] Burciaga-Diaz O, Escalante-Garcia JI, Gorokhovsky A. Geopolymers based on a coarse

low-purity kaolin mineral: Mechanical strength as a function of the chemical composition and temperature. Cem Concr Compos 2012; 34: 18-24.

[36] Michelina Catauroa, Ferdinando Papalea, Giuseppe Lamannaa, et al. Synthesis and Investigation of the Polymer Influence on Microstructure and Mechanical Behavior. Mater Res 2014; 18: 698-705.

[37] Shadnia R, Zhang L, Li P. Experimental study of geopolymer mortar with incorporated PCM. Constr Build Mater 2015; 84: 95-102.

[38] Cao VD, Pilehvar S, Salas-Bringas C, et al. Influence of microcapsule size and shell polarity on thermal and mechanical properties of thermoregulating geopolymer concrete for passive building applications. Energy Convers Manag 2018; 164: 198-209.

[39] Pilehvar S, Cao VD, Szczotok AM, et al. Mechanical properties and microscale changes of geopolymer concrete and Portland cement concrete containing micro-encapsulated phase change materials. Cem Concr Res 2017; 100: 341-349.

[40] Frattini D, Roviello G, Ferone C, et al. Geopolymer-Based Composite Materials Containing Pcm for Thermal Energy Storage. Ceram Energy, Naples, Italy, *2015* 2015; 6-7.

[41] Sukontasukkul P, Nontiyutsirikul N, Songpiriyakij S, et al. Use of phase change material to improve thermal properties of lightweight geopolymer panel. Mater Struct Constr 2016; 49: 4637-4645.

[42] Wang Z, Su H, Zhao S, et al. Influence of phase change material on mechanical and thermal properties of clay geopolymer mortar. Constr Build Mater 2016; 120: 329-334.

[43] Kastiukas G, Zhou X, Castro-Gomes J. Development and optimisation of phase change material-impregnated lightweight aggregates for geopolymer composites made from aluminosilicate rich mud and milled glass powder. Constr Build Mater 2016; 110: 201-210.

[44] Razak RA, Khang Zhe AC, Bakri Abdullah MM Al, et al. Paraffin as a Phase Change Material in Concrete for Enhancing Thermal Energy Storage. *IOP Conf Ser Mater Sci Eng*; 743. Epub ahead of print 2020. DOI: 10.1088/1757-899X/743/1/012012.

[45] Hunger M, Entrop AG, Mandilaras I, et al. The behavior of self-compacting concrete containing micro-encapsulated Phase Change Materials. Cem Concr Compos 2009; 31: 731-743.

[46] Sari A, Biçer A. Preparation and thermal energy storage properties of building material-based composites as novel form-stable PCMs. Energy Build 2012; 51: 73-83.

[47] Craveiro F, Nazarian S, Bartolo H, et al. An automated system for 3D printing functionally graded concrete-based materials. Epub ahead of print 2020. DOI: 10.1016/j.addma.2020.101146.

[48] Vlachakis C, Perry M, Biondi L, et al. 3D printed temperature-sensing repairs for concrete structures. Addit Manuf 2019; 101238.

[49] Khan MA. Mix suitable for concrete 3D printing : A review. *Mater Today Proc*. Epub ahead of print 2020. DOI: 10.1016/j.matpr.2020.03.825.

[50] Xia M, Sanjayan J. Method of formulating geopolymer for 3D printing for construction applications. Mater Des 2016; 110: 382-390.

[51] Wu P, Wang J, Wang X. A critical review of the use of 3-D printing in the construction industry. Autom Constr 2016; 68: 21-31.

[52] Khalil N, Aouad G, El K, et al. Use of calcium sulfoaluminate cements for setting control of 3D-printing mortars. Constr Build Mater 2017; 157: 382-391.

[53] Singh S, Aswath MU, Ranganath R V. Effect of mechanical activation of red mud on the strength of geopolymer binder. Constr Build Mater 2018; 177: 91-101.

[54] Yamchelou MT, Law DW, Patnaikuni I, et al. Alkali activation of mechanically activated low-grade clay. J Sustain Cem Mater 2020; 1-17.

[55] Al Bakri Abdullah MM, Hussin K, Bnhussain M, et al. Fly ash-based geopolymer lightweight concrete using foaming agent. Int J Mol Sci 2012; 13: 7186-7198.

[56] Sukontasukkul P, Chindaprasirt P, Pongsopha P, et al. Effect of fly ash/ silica fume ratio and curing condition on mechanical properties of fiber-reinforced geopolymer. J Sustain Cem Mater 2020; 0: 1-15.

[57] Ma G, Li Z, Wang L, et al. Micro-cable reinforced geopolymer composite for extrusion-based 3D printing. Mater Lett 2019; 235: 144-147.

[58] Shakor P, Sanjayan J, Nazari A, et al. Modified 3D printed powder to cement-based material and mechanical properties of cement scaffold used in 3D printing. Constr Build Mater 2017; 138: 398-409.

[59] Ibrahim M, Biwole PH, Wurtz E, et al. A study on the thermal performance of exterior walls covered with a recently patented silica-aerogel-based insulating coating. Build Environ 2014; 81: 112-122.

[60] Yu CJ, Ri BH, Kim CH, et al. Formation and characterization of ceramic coating from alumino silicate mineral powders in the matrix of cement composite on the concrete wall. Mater Chem Phys 2019; 227: 211-218.

[61] Zhao J, Liebscher M, Michel A, et al. Plasma-generated silicon oxide coatings of carbon fibres for improved bonding to mineral-based impregnation materials and concrete matrices. Cem Concr Compos 2020; 103667.

[62] Sow M, Hot J, Tribout C, et al. Characterization of Spreader Stoker Coal Fly Ashes (SSCFA) for their use in cement-based applications. Fuel 2015; 162: 224-233.

[63] Mohseni E, Tang W, Khayat KH, et al. Thermal performance and corrosion resistance of structural-functional concrete made with inorganic PCM. Constr Build Mater 2020; 249: 118768.

[64] Karlessi T, Santamouris M, Synnefa A, et al. Development and testing of PCM doped cool colored coatings to mitigate urban heat island and cool buildings. Build Environ 2011; 46: 570-576.

[65] Ferrer G, Solé A, Barreneche C, et al. Corrosion of metal containers for use in PCM energy storage. Renew Energy 2015; 76: 465-469.

[66] Zainal FF, Fazill MF, Hussin K, et al. Effect of Geopolymer Coating on Mild Steel. 2018; 273: 175-180.

[67] Zheng W, Chen WG, Feng T, et al. Progress in Organic Coatings Enhancing chloride ion penetration resistance into concrete by using graphene oxide reinforced waterborne epoxy coating. 138. Epub ahead of print 2020. DOI: 10.1016/j.porgcoat.2019.105389.

[68] Zheng W, Chen WG, Feng T, et al. Enhancing chloride ion penetration resistance into concrete by using graphene oxide reinforced waterborne epoxy coating. *Prog Org Coatings*; 138. Epub ahead of print 2020. DOI: 10.1016/j.porgcoat.2019.105389.

[69] Zhong Z, Sha Q, Zheng J, et al. Sector-based VOCs emission factors and

source profiles for the surface coating industry in the Pearl River Delta region of China. Sci Total Environ 2017; 583: 19-28.

[70] Cheng L, Liu C, Han D, et al. Effect of graphene on corrosion resistance of waterborne inorganic zinc-rich coatings. J Alloys Compd 2019; 774: 255-264.

[71] Cheng L, Liu C, Han D, et al. Effect of graphene on corrosion resistance of waterborne inorganic zinc-rich coatings. J Alloys Compd 2019; 774: 255-264.

[72] Rieger D. Corrosion Studies of Fly Ash and Fly Ash-Slag Based. DOI: 10.1088/1757-899X/209/1/012026.

[73] Afshar A, Jahandari S, Rasekh H, et al. Corrosion resistance evaluation of rebars with various primers and coatings in concrete modified with different additives. Constr Build Mater 2020; 262: 120034.

[74] Zhao Z, Qu X, Li J. Application of polymer modified cementitious coatings (PCCs) for impermeability enhancement of concrete. Constr Build Mater 2020; 249: 118769.

[75] Xu K, Ren S, Song J, et al. Colorful superhydrophobic concrete coating. Chem Eng J 2021; 403: 126348.

[76] Morefield SW, Hock VF, Weiss CA, et al. STEEL IN CEMENT-BASED COMPOSITES.

[77] Karaipekli A, Sari A. Capric-myristic acid/expanded perlite composite as form-stable phase change material for latent heat thermal energy storage. Renew Energy 2008; 33: 2599-2605.

[78] Karaipekli A, Sari A. Capric-myristic acid/vermiculite composite as form-stable phase change material for thermal energy storage. Sol Energy 2009; 83: 323-332.

[79] Sari A. Fabrication and thermal characterization of kaolin-based composite phase change materials for latent heat storage in buildings. Energy Build 2015; 96: 193-200.

[80] Yuan Y, Zhang N, Tao W, et al. Fatty acids as phase change materials: A review. Renew Sustain Energy Rev 2014; 29: 482-498.

[81] Zhang N, Yuan Y, Yuan Y, et al. Lauric-palmitic-stearic acid/expanded perlite composite as form-stable phase change material: Preparation and thermal properties. Energy Build 2014; 82: 505-511.

[82] Lv P, Liu C, Rao Z. Review on clay mineral-based form-stable phase change materials: Preparation, characterization and applications. Renew Sustain Energy Rev 2017; 68: 707-726.

[83] Rashidi S, Esfahani JA, Karimi N. Porous materials in building energy technologies—A review of the applications, modelling and experiments. Renew Sustain Energy Rev 2018; 91: 229-247.

Section 2

Zeolites

Chapter 6

Genesis of Structure and Properties of the Zeolite-Like Cement Matrices of the System Na(K)-Al_2O_3-SiO_2-H_2O within a Temperature Range of 20–1200°C

Pavel Krivenko and Volodymyr Kyrychok

Abstract

The chapter deals with a genesis of structure and properties of the zeolite-like cement matrices of the Na(K)-Al_2O_3-SiO_2-H_2O system within a temperature range of 20–1200°C. Due to the fact that zeolite-like structures and their characteristics vary within wide ranges, materials with high-performance properties can be obtained through regulation of the structure formation processes. This can be provided by a proper choice of type of an aluminosilicate component, cation of an alkaline component and additives, including Ca-containing ones, and curing conditions. When the cement matrix formation process is appropriately directed, the zeolite-like products (hydrosodalite, analcime, chabasite, faujasite etc.) dominate in the microstructure that is formed. The ability of some zeolites to recrystallize with temperature increase into stable feldspar-like aluminosilicates without destroying the basic skeleton opens a pathway that is worth to explore in the production of materials similar to low temperature ceramics, intumescent coatings, high temperature and corrosion resistant structures, etc. The examples are given on how to use the above cement matrices for making some of the above listed materials.

Keywords: alkali, aluminosilicate, cation, fly ash, metakaolin, phase transformation, pore, strength, thermal shrinkage, zeolite phases

1. Introduction

Technical advances in building industry in association with progress in knowledge collected on the chemistry of cementitious materials, requirements of global sustainable development serve as a moving force in further development of the alkali-activated materials [1–3].

Earlier, according to traditional views, free alkalis and compounds of alkali metals were excluded from traditional hydraulic cement compositions because of their high solubility which resulted in the worser durability and properties.

However, the studies held on ancient cements and concretes showed that their excellent durability could be attributed to the presence of aluminosilicates of alkali metals – analogs of natural zeolites – in their hydration products. Alkalis play an important role in mutual transformations of minerals in the Earth's crust. It should be also mentioned that aluminosilicates of alkali metals (first of all, feldspars) are known to be more stable and durable compared to those of calcium [1].

In 1957 Viktor Glukhovsky made an attempt to model natural processes of formation of the aluminosilicates of alkali metals at different temperatures and made an assumption that the compounds of alkali metals (Li, Na, K, Rb, Cs) – the elements of Group 1 of the periodic table of elements, exhibited hydraulic binding properties similar to compounds of the alkaline earth metals (Mg, Ca, Sr, Ba) – the elements of Group 2 [4].

An important conclusion was also made that the increase of temperature promoted a smooth dehydration process and subsequent re-crystallization of the hydration products into stable anhydrous aluminosilicates of alkali metals. This conclusion was later confirmed [5–7]. Unique service properties of materials predetermine its application for development of a wide area of modern composite materials for the building industry such as protective coatings, inorganic glues, heat-insulating materials, thermo resistance composites, etc. [8–14].

Extensive works carried out have allowed to revise views of regularities of exhibiting binding properties by mineral substances and to prove that, in parallel with the compounds formed by the elements of Group 2 (alkaline earth elements) of the periodic table of elements together with complex-formers of Groups 3, 4, 5 or 6, the compounds formed by elements of the main Subgroup of Group 1 together with twin complex-formers of the Groups 3 and 4 (aluminosilicates) possess binding properties. Such products can be formed through a combination of other amphoteric and acid compounds [4, 8, 9].

The established regularities governing the formation of a mineral-like cement matrix based on the compounds of alkali metals taken alone and in combination with the compounds of alkaline earth metals were established and studied in details (hardening processes, principles of compositional structure and prediction of their properties) and used for development of a new class of hydraulic binders known nowadays under a name of alkali-activated cements or alkali-activated aluminosilicate cements [10–17].

The idea behind the alkali activated cements is modeling of the minerals of the Earth's crust in the system of Me_2O-Me_2O_3-SiO_2-H_2O (Me_2O- Li_2O, Na_2O, K_2O, Rb_2O, Cs_2O; MeO-CaO, MgO; Me_2O_3-Al_2O_3, Fe_2O_3, Cr_2O_3). For example, aluminosilicate rocks of feldspar composition are broken up by chemical wind erosion to a dispersion state or to clay minerals. These transformations are accompanied by hydration of anhydrous minerals composed from the compounds of alkali metals, the decrease of alkali content in hydrated new formations; the replacement of alkalis by hydrogenous ions or H_3O^+-groups; the transition of aluminium from IV-coordination to VI-one; the partial removal of silicic acid, i.e. by processes occurring under hydration and hardening of building cements. Gel-like silicate and aluminosilicate substances are formed as a result of chemical erosion of feldspars in an erosion crust, in the zones of hydrothermal metamorphism proceeding at relatively low temperatures and pressures. These substances react with compounds of alkali metals brought by circulating superheated aqueous solutions. As a result, aluminosilicate hydrates of alkali metals of zeolite type are formed. These minerals have a very low water solubility despite the fact that their composition includes strongly soluble compounds of alkali metals [5–8].

In particular, analcime ($Na_2O{\cdot}Al_2O_3{\cdot}4SiO_2{\cdot}2H_2O$) is formed at 303 K (29.85°C) at the sea bed through a coagulation of silica and alumina sols with their concurrent adsorption of ions og alkali metals from environment.

Geological data suggest that on numerous occasions aluminosilicates of alkali metals of zeolite type or sodalite, feldspathoid, feldspar types (these minerals are structurally analogous to zeolites) which are the rock-forming minerals, were formed at the expense of chemical interaction between a clayey substance and compounds of alkali metals in an erosion burst. But these processes in nature proceed very slowly, during geological periods. The presence of an alkaline medium is the determining factor for the progress of these processes.

The processes in natural or compositionally analogous to natural, artificial aluminosilicates of alkali metals can be accelerated to limits wherein aluminosilicates can be used as hydraulic binders. Similar to the production of Portland cement clinker, it is possible through the conversion of these substances from the stable crystalline state into a more active metastable one, including a glassy state, or through the external introduction of compounds of alkali metals [5].

Extensive experimental studies confirmed the proposed theoretical bases and allowed to prove that caustic alkalis; salts of alkali metals and weak acids; silicates, aluminates and aluminosilicates reacting in the presence of alkalis under condition that their concentration in the system is sufficient react with natural clay minerals or undead burnt ones; with natural and artificial aluminosilicate glasses, among them metallurgical slags, fuel ashes and slags. The processes can take place in natural conditions and under steam curing and a water-resistant stone with the hydration products that are analogous to natural minerals of zeolite type is formed [3, 4, 8, 9].

Main criteria to be applicable to a choice of starting materials for the zeolite-like cement matrices are: high contents of silica and alumina in them which promote a synthesis of the zeolite-like phases of the $(Na, K)_2O$-Al_2O_3-SiO_2-H_2O system and some content of calcium oxide which will serve as an inhibitor of the process of zeolite synthesis [7–9]. The most appropriate for these goals are natural clays in natural and dehydrated state, volcanic rocks or industrial by-products like fuel ashes and slags, metallurgical slags, red muds etc. [3, 4, 9, 17, 18].

Large differences can be observed between microstructures of these cement matrices. A microstructure of the metakaolin derived cement stone, for example, was investigated by systematic variation of activator composition and related to mechanical strength [18–31]. The most important factors affecting a phase composition and, correspondingly, properties of the material are: curing conditions and type and concentration of the alkaline activator solution (that is, the ratios between main oxides (Na_2O/Al_2O_3 and SiO_2/Al_2O_3) as well as a water to solid (W/S) ratio [4, 9, 11, 18–23, 30, 31]. It was observed that with the Si/Al ratio increase, a porosity of the microstructure changed from large pores to more homogenous structure with small pores. This observation was linked to a strong correlation with the Young's modulus and mechanical strength increase [22].

The majority of researchers [32–34] made a conclusion that the structure formation processes in the alkali activated aluminosilicate systems were determined by a required constitutional composition and flowed step-by-step with the formation, depending on a temperature, of the amorphous, glassy or crystalline zeolite-like cement matrices.

For example, reaction products of the interaction of clay minerals with NaOH at 100–300°C are hydronephelines with a structure of sodalite, and those of the interaction with KOH – caliophilites containing zeolitic and adsorbed water.

Hydronepheline and caliophilite themselves can act in a ceramic matrix as structure forming binders.

Low-temperature ceramics produced from them is characteristic of the enhanced resistance to alkalis. With further curing temperature increase of the mineral system to be synthesized the process will flow under a dehydration-condensation scheme. Hydrates that are formed at low temperatures start to dehydrate, their crystal lattices destroy and the substance becomes amorphous. Further transformations of the dehydration products are attributed, in case of the sodium compounds, to their transformation at 100–800°C into nepheline, at 800–1100°C into albite under condition of required amount of silica available in the system, in case of the potassium compounds – first, into anhydrous caliophilite, afterwards – into orthoclase. The formation of a hydrate phase as a result of the interaction of minerals of unfired clays with salts of alkali metals in normal conditions is rather difficult to identify. However, being subjected to heating, they transform into anhydrous minerals of nepheline, leucite, afterwards – into albite and orthoclase. Their synthesis is preceded by the processes of dehydration of the clay minerals and thermal dissociation of the alkali metal salts.

A character and temperature at which the products are synthesized is determined by a structural type of a clay mineral, type of compound of alkali metal, and nature of siliceous component.

The phases composed from aluminosilicates of alkali metals of the albite and orthoclase type are the most intensively crystallized in the compositions which contain bentonite, and with the lowest intensity in the compositions which contain kaolin. The additives of mineralizers, amorphous silica, α-cristobalite and preliminary dehydration of the clays render an intensifying action on crystallization of these phases.

In the mixed binding systems of the $Na_2O(CaO)$-Al_2O_3-SiO_2-H_2O composition the structure formation processes are determined by crystallization of the phases which are analogs to natural plagioclases.

A purpose of the paper is to make a systematic analysis of transformation of phase and properties of the zeolite-like cement matrices of the system Na(K)-Al_2O_3-SiO_2-H_2O in relation to various factors, the most important of them are: temperature of curing, reactive silica content (SiO_2/Al_2O_3 ratio), initial alkalinity (Na_2O/Al_2O_3 ratio), cation type (Na or K, and Ca), type of aluminosilicate component (metakaolin or fuel ash (fly ash).

2. Experimental

The cement compositions used in the study were formulated from an aluminosilicate component selected from a typical metakaolin and two type F fly ashes ($380 m^2/kg$ by Blaine) (**Table 1**) and soluble sodium silicates (M = 2.8 and density = 1400 kg/m^3) as an alkaline component. In some cases sodium hydroxide and silica fume were added as supplementary components. In the experiments on thermal behaviour exclusively in the tests held at high temperatures, a ground chamotte with a specific surface area = 410 m^2/kg by Blaine was used as filler. After curing in an air tight mould at 80°C until solidification, the specimens were then were subjected for 6 hr- curing: in autoclave at 174°C, in steam curing chamber at 80°C, and in drying chamber at 80, 150, and 220°C. The structure formation processes at

Constituent	**Chemical composition (% by mass)**											**In total**
	SiO_2	**TiO_2**	**Al_2O_3**	**Fe_2O_3**	**FeO**	**MnO**	**MgO**	**CaO**	**R_2O**	**P_2O_5**	**LOI**	
Metakaolin	55.76	0.00	35.86	4.33	0.00	0.00	0.93	3.05	0.00	0.00	0.07	100.00
Fly ash 1	50.64	1.21	23.73	9.67	0.71	0.15	1.76	2.32	3.26	0.45	6.10	100.00
Fly ash 2	48.25	0.89	19.67	4.50	3.15	0.11	1.36	2.18	3.82	0.02	16.04	100.00
Silica fume	87.58	0.00	0.20	0.14	0.00	0.00	0.50	2.10	4.76	0.00	4.71	100.00

Table 1.
Chemical composition of raw materials.

high temperatures were studied on the specimens that then additionally cured for 4 hrs within a temperature interval of 600–1200°C. An interrelation between mechanical properties determined on the specimens of size 2 × 2 × 2 cm and the results of examination of a phase composition studied by means of XRD, DTA, and SEM techniques was established.

On the contrary to a traditional approach, under which changes in a pH-value and an aluminosilicate component to alkaline solution ratio were taken as variables, an approach applied in this study was based on a ratio between the main constituent oxides in a reacting mix (mainly, SiO_2/Al_2O_3); it makes more easier to understand the structure formation processes with regard to peculiarities of a synthesized microstructure.

3. Results

3.1 The structure formation processes in the cement of the R_2O-Al_2O_3-SiO_2-H_2O system

3.1.1 Influence of curing conditions

Ten cement compositions were chosen for this study (**Table 2**). Each cement composition was cured using one out of five curing regimes mentioned above. The XRD and DTA results allowed identifying the fields of crystallization of the aluminosilicate-based hydration products (**Figure 1**).

Thus, the phase compositions of the reaction products of the metakaolin- and fly ash-based cements are similar. The phase composition of the reaction products of the autoclaved fly ash-based cements is represented mainly by analcime, zeolite-P and zeolite-R. These conclusions coincide well the data obtained on the model systems and reported in [5, 6] and those obtained with the metakaolin-based cements [12].

A synthesis of hydroxysodalite which is formed due to high contents of Na-cation in the steam cured fly ash-based cement compositions with SiO_2/Al_2O_3 = 4 and those after drying correlates well with the data obtained on similar systems with high alkali concentrations and after curing in normal conditions [5, 6, 35–37]. When using the low-alkali cement compositions with SiO_2/Al_2O_3 = 6–8

Aluminosilicate component	Mix No	Molar ratio		
		Na_2O/Al_2O_3	SiO_2/Al_2O_3	H_2O/Al_2O_3
Metakaolin	1	1.0	2.0	9.5
	2	1.0	4.0	10.0
	3	1.0	6.0	15.0
	4	1.0	8.0	20.0
Fly ash 1	5	1.0	4.0	7.0
	6	1.0	6.0	10.5
	7	1.0	8.0	16.5
Fly ash 2	8	1.0	4.0	10.0
	9	1.0	6.0	11.5
	10	1.0	8.0	17.0

Table 2.
Ratio of the components in a cement matrix.

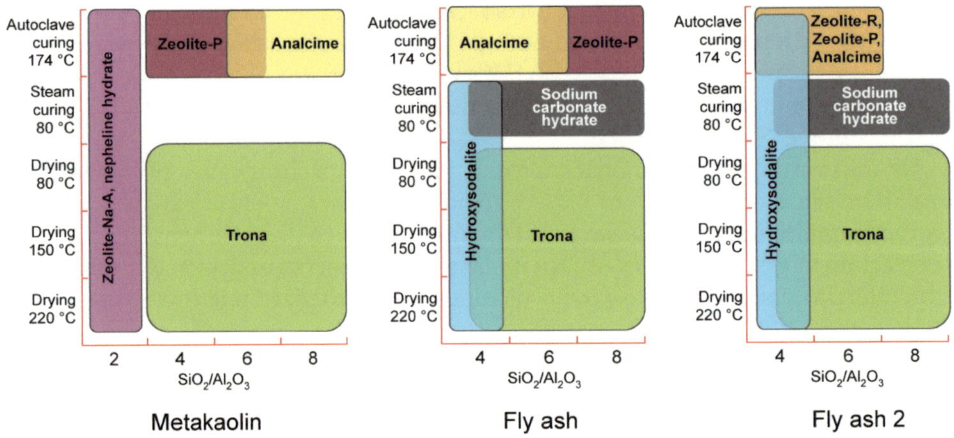

Figure 1.
Type of hydration products vs. aluminosilicate component of the cements.

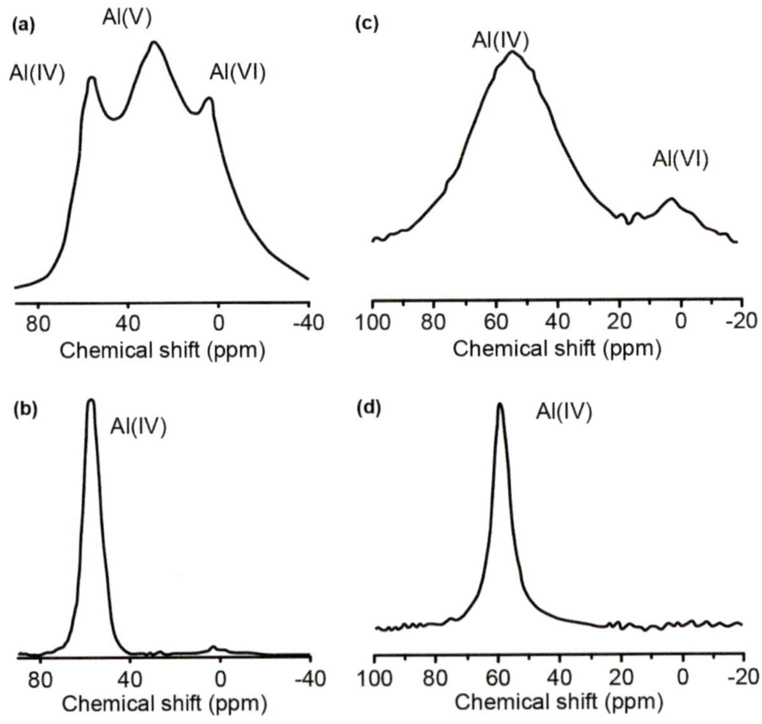

Figure 2.
The ^{27}Al MAS-NMR spectra of (a) typical metakaolin, (b) metakaolin-based Na-geopolymer from metakaolin with Si/Al ratio of 2.15, (c) typical fly ash, and (d) Na-geopolymer from NaOH activation of fly ash (20 hours, 85°C) [38].

simultaneously with a low-temperature curing such as steam curing or drying, the accepted exposition time is not enough for the zeolite crystallization.

A speed of the mineral (in our case, zeolite) formation is usually measured by a height of the XRD peaks. These peaks in case of the fly ash-based cements are much more higher compared to those in case of the metakaolin-based cements.

This also follows from the Si MAS-NMR and the Al MAS-NMR spectra (**Figures 2** and **3**) [38–46].

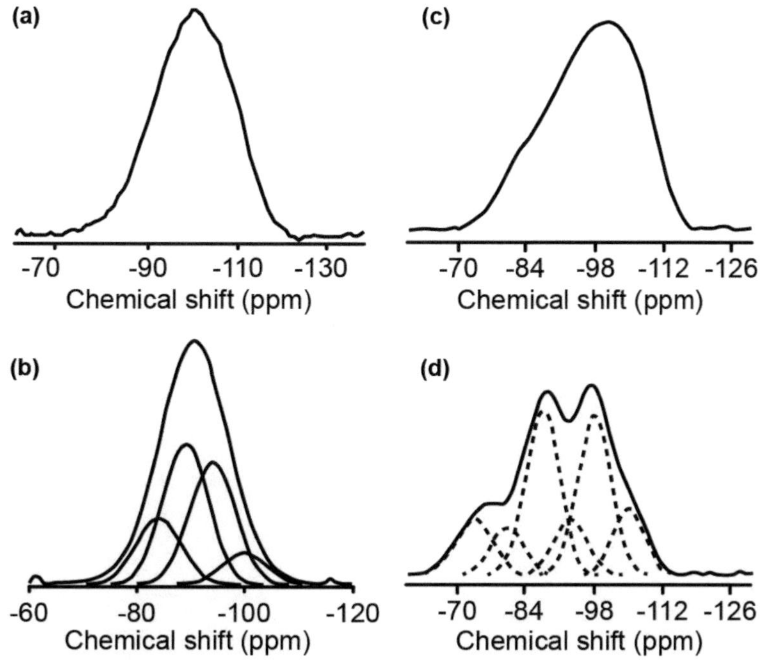

Figure 3.
The ^{29}Si MAS-NMR spectra of (a) typical metakaolin, (b) metakaolin-based Na-geopolymer with Si/Al ratio of 1.65, (c) typical fly ash, and (d) fly ash-based Na-geopolymer (20 hours, 85°C) [39].

The greater intensively of the peaks 93 and 98 ppm in the fly ash-based cement compared to the signal at 84.6 and 89 ppm in the metakaolin-based cement is an indication that the aluminosilicate gel that is formed (N-A-S-H) in the fly-ash based cement has a higher silica content.

After drying of both metakaolin- and fly ash-based cements, a minor phase was fixed at first as a "phase Z" [18], but further studies on carbonation finally showed that these peaks could be identified as a trona mineral ($Na_2CO_3 \cdot NaHCO_3 \cdot 2H_2O$). After steam curing, another minor phase with main peaks (0.276, 0.237 nm) was identified [18]. These peaks can be an evidence of formation of another sodium carbonate, $Na_2CO_3 \cdot H_2O$ (due to carbonation process). These two phases may sometimes co-exist, however, their formation takes place under a different mechanism: the higher W/S ratios, the higher amounts of the trona are formed. A synthesis of the sodium carbonate hydrate takes place only at low W/S ratios and high alkalinity under hydrothermal conditions (steam curing, curing in an air tight mould).

In order to study a phase composition of dehydration products the cement specimens (**Table 2**) were, after pre-curing, subjected to high temperature curing (800°C). On the contrary to the phase composition after hydration, which is dependent, above all, on a mode of pre-curing, the phase composition after dehydration depends only on the cement composition and does not depend on the mode of pre-curing. So, in the cements with SiO_2/Al_2O_3 = 4 a phase composition of the reaction product is represented by nepheline (**Figure 4**). The degree of crystallization does not depend on the mode of pre-curing, and depends on a type of the aluminosilicate component: a maximal degree of crystallization was observed in case of the metakaolin compared to the fly ashes. This can be attributed to the fact that the metakaolin is the most chemically pure component. The impurities from the fly ashes resulted in the reduced contents of the anhydrous aluminosilicates and in the formation of additional amount of hematite.

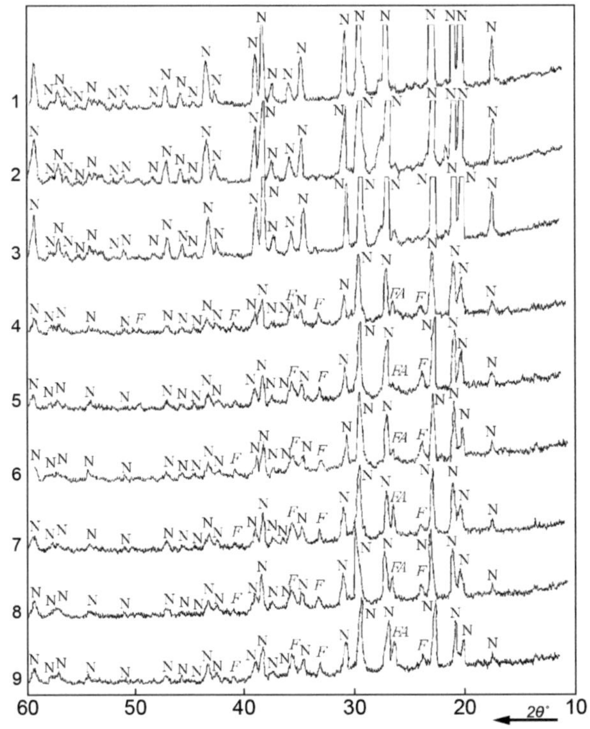

Figure 4.
The XRD scans of the metakaolin-based cements with SiO_2/Al_2O_3 = 4 after high temperature curing at 800°C. Aluminosilicate component: Metakaolin (1–3), fly ash 1 (4–6), fly ash 2 (7–9). Pre-curing mode: Steam curing (1, 4, 7), drying (2, 5, 8), without pre-curing (3, 6, 9) N – Nepheline, Ab – Albite, Kr – Cristobalite, F – Hematite, FA – Residue of unreacted fly ash.

3.1.2 The role of type of the aluminosilicate component and SiO_2/Al_2O_3 ratio

Figures 5 and **6** demonstrate XRD scans and SEM images of the microstructure of a cleaved fragment of the artificial stone depending on the ratio SiO_2/Al_2O_3 = 2:7 in the original composition of the zeolite-like cement matrices after drying at 80°C (under condition that K_2O/R_2O = 0.15 and $(Na_2O + K_2O)/Al_2O_3$ = 1).

A phase composition of the hydration products at low of SiO_2/Al_2O_3 ratios (2:3) after drying at 80°C is represented by zeolite-Na-A (d/n = 0.699; 0.365; 0.336; 0.293 nm), natrolite (d/n = 0.287; 0.243; 0.138 nm), ussingite (d/n = 0.492; 0.347; 0.295 nm). Amorphous phases of the metakaolin-based cement and particles of the non-reacted metakaolin can be clearly seen in the SEM images (**Figure 6**).

The alkali-activated aluminosilicate cements with the SiO_2/Al_2O_3 ratios =4:5 have the zeolite-like hydration products of the following type: zeolite-Na-A (d/n = 0.699; 0.365; 0.336; 0.293 nm), heulandite-Na (d/n = 0.509; 0.392; 0.296 nm), heulandite-K (d/n = 0.342; 0.281; 0.273 nm), and phillipsite-Na-K (d/n = 0.498; 0.408; 0.269 nm).

As it follows from **Figures 7** and **8**, with the temperature increase the degree of crystallinity increases as well.

The 3D-polymeric crystalline or semi-crystalline aluminosilicates of alkali metals are main hydration products of the cements under study at low temperatures (usually below 200°C). Correspondingly, they refer to "zeolites" or "zeolite pre-cursors". Thermal stability of these phases depends on their structure: some zeolite structures are known to be resistant to heating, and some of them decompose in the process of heating. Among various zeolites able to crystallize in the cement matrix,

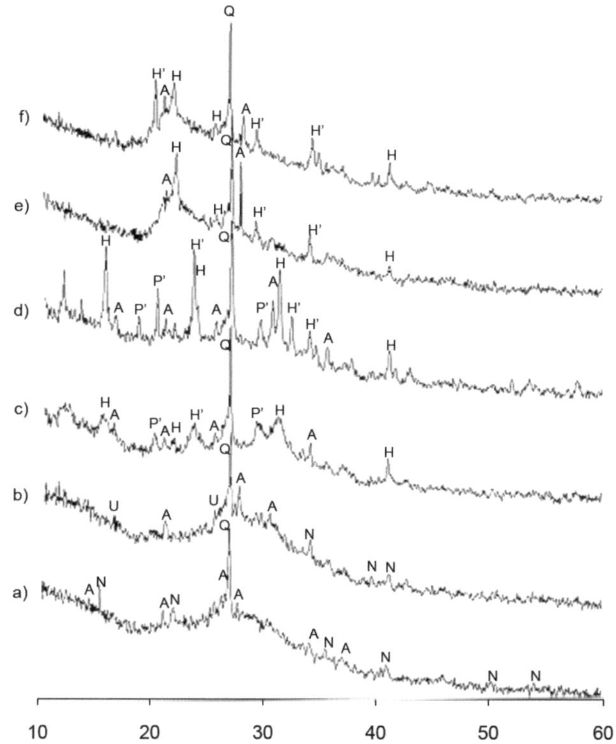

Figure 5.
The XRD scans of the metakaolin-based cements after drying at 80°C (K_2O/R_2O = 0.15 and SiO_2/Al_2O_3, respectively: a – 2; b – 3; c – 4; d – 5; e – 6; f – 7. Q – Quartz; N – Natrolite; a – Zeolite-Na-a; U – Ussingite; P′ – Phillipsite-Na-K; H – Heulandite-Na; H′ – Heulandite-K.

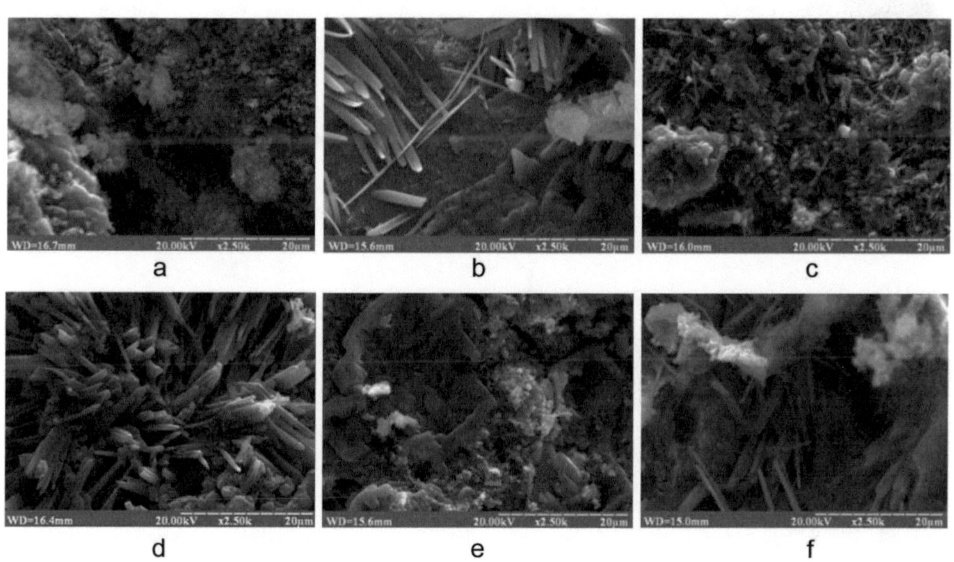

Figure 6.
The SEM images of the metakaolin-based cements after drying at 80°C (K_2O/R_2O = 0.15 and SiO_2/Al_2O_3, respectively: a – 2; b – 3; c – 4; d – 5; e – 6; f – 7 (×2500).

only certain zeolites remain their structure until 700–800°C. Thermally stable structures are: sodalite network (hydroxysodalite), analcime, chabazite structures (zeolite-R, herchelite), faujasite family (zeolites-Na-X and Na-Y), mordenite [5–7].

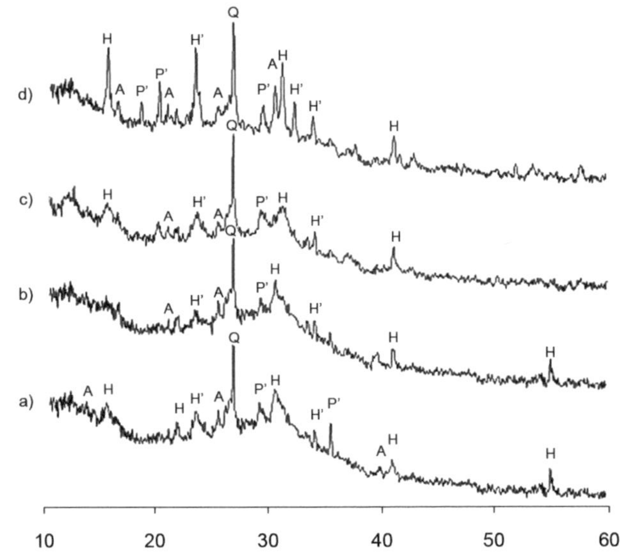

Figure 7.
The XRD scans of the metakaolin-based cement (K_2O/R_2O = 0.15 and SiO_2/Al_2O_3 = 5) after drying at temperatures, °C: a – 20; b – 40; c – 60; d – 80. Legend: Q – Quartz; A – Zeolite Na-A; P' – Phillipsite-Na-K; H – Heulandite-Na; H' – Heulandite-K.

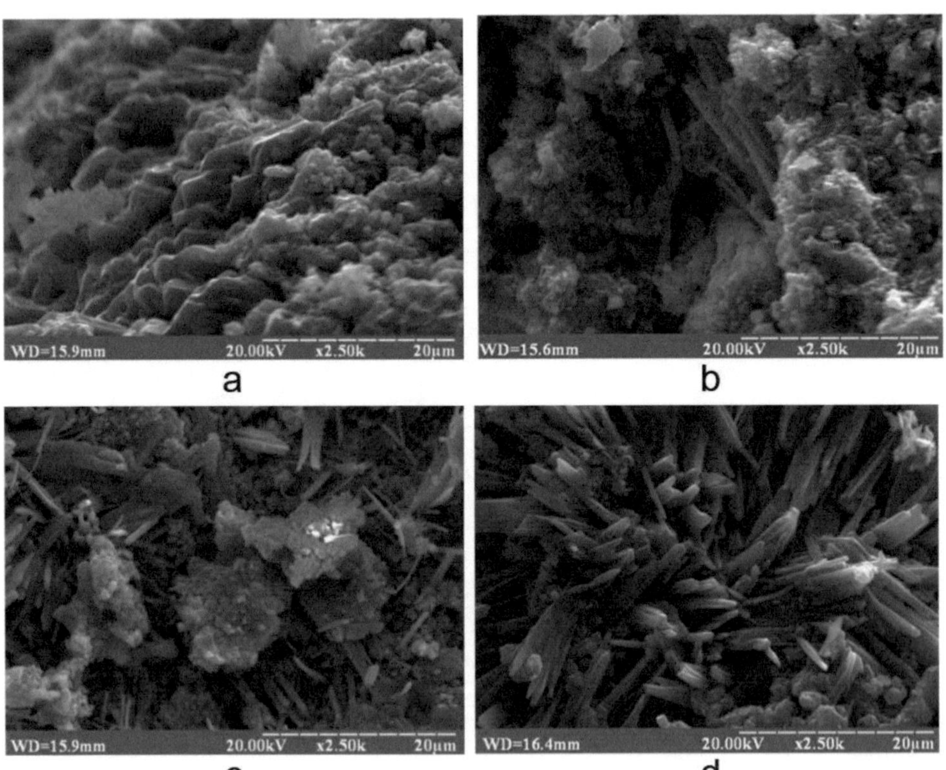

Figure 8.
The SEM images of the microstructure of the metakaolin-based cement (K_2O/R_2O = 0.15 and SiO_2/Al_2O_3 = 5) after drying at temperatures, °C: a – 20; b – 40; c – 60; d – 80 (×2500).

Once these phases (or N-A-S-H gels of similar structure) are synthesized in the cement matrix, they slightly improve their crystallinity during heating until 200–400°C, and then keep their structure up to approx. 800°C and after this re-crystallize with the formation of structurally similar nepheline or albite. Other zeolitic structures, for example, zeolites-Na-A or P, decompose at temperatures within the range of 120–300°C, making impossible to use them in the materials that are resistant at high temperatures.

In studying the influence of the cement composition on the phase composition after heating above the temperature of dehydration, a conclusion was drawn on a correlation between the SiO_2/Al_2O_3 molar ratio in the initial cement compositions and their dehydration products (**Table 3**). With increase of the ratio from 2 to 8, the reaction products transform in the following direction: nepheline ($Na_2O \cdot Al_2O_3 \cdot 2SiO_2$) – albite ($Na_2O \cdot Al_2O_3 \cdot 6SiO_2$) – α-cristobalite ($SiO_2$) (**Figure 4**). This tendency in general coincides well with the data obtained on the model systems of the alkali-activated aluminosilicate cements [18, 47]. So, a feldspathoid nepheline (SiO_2/Al_2O_3 = 2 in the aluminosilicate framework) appears in a phase composition of the cements with SiO_2/Al_2O_3 = 2–4, irrespectively of a type of the aluminosilicate component used. A maximum XRD- measured quantity of nepheline was found in case when a SiO_2/Al_2O_3 ratio was 2. The increase of the ratio up to 6–8 resulted in the additional formation of α-cristobalite (d = 0.411; 0.252; 0.206; 0.163 nm). The higher contents of amorphous silica, the higher speed of α-cristobalite formation.

In the fly ash-based cements with the SiO_2/Al_2O_3 ratio = 6, the reaction products are represented by high-silica albite and some amounts of nepheline (**Table 3**). Differences in a phase composition of the metakaolin-based and fly-ash-based cements can be attributed to high contents of main oxides in a non-active form such as quartz and mullite which reduce a real stoichiometric ratio in these two cements.

As mentioned above, a type of the aluminosilicate component used (metakaolin or fly ash) does not affect the differences: once heated to high temperatures (600–1200° C), a microstructure of the cements under study tends to transform into that with the phases resistant to high temperatures. In contrast to the OPC-based cements, no any instable phase was identified. However, metakaolin, a purer aluminosilicate component, gives the higher purity and higher content of the formed high-temperature phases formed. In case of the fly ash the lower quantities of the zeolite-like products, however, with a more variable composition due to small quantities of such oxides as Fe_2O_3, MgO, CaO, TiO_2 etc. contained in the fly ash are formed.

Aluminosilicate component	SiO_2/Al_2O_3 molar ratio	
	Initial cement	Main phase
Metakaolin	2	**nepheline**
	4	**nepheline**
	6	**α-cristobalite**, nepheline
	8	**α-cristobalite**, nepheline
Fly ash 1	4	**nepheline**
	6	**α-cristobalite**, nepheline
	8	**α-cristobalite**, nepheline

Remark: Dominating phases are given in bold.

Table 3.
The SiO_2/Al_2O_3 molar ratio of the cements under study after high temperature curing (1200°C).

3.1.3 Role of initial alkalinity (Na_2O/Al_2O_3 ratio)

The alkalis in the alkali-activated aluminosilicate cements participate in all stages of structure formation. In particular, rate and degree of activation of the aluminosilicate component is determined by contents of alkalis [3, 31, 36]. A quantity of alkalis is to be chosen in view of formation of a required phase composition which determines the required properties. For example, the anhydrous aluminosilicates appear in small amounts in the systems with high content of alkali (Na_2O/Al_2O_3 = 1.0) at 600°C, whereas in the low-alkali compositions (Na_2O/Al_2O_3 = 0.5) the crystallization begins only at 1000°C (**Table 4**) [36, 37].

3.1.4 Role of a cation type and direction of phase transformations

Figures 9 and **10** demonstrate the XRD scans and SEM images of the microstructure of a cleaved fragment of the artificial stone (K_2O/R_2O = 0:0.3, $(Na_2O + K_2O)/Al_2O_3$ = 1 and SiO_2/Al_2O_3 = 5) after curing at 80°C. At the ratio SiO_2/Al_2O_3 = 5, a phase composition of the artificial stone is characterized by the zeolite-like hydration products of the heulandite and phillipsite types (**Figure 9**), and the

No	Main oxides, molar ratio			Phase composition after curing at			
	SiO_2/Al_2O_3	Na_2O/Al_2O_3	Na_2O (%)	600°C	800°C	1000°C	1200°C
1	4.0	0.5	6.5	—	N (low)	N, A	A
2	4.0	1.0	12.9	—	N	N	(g)
3	3.5	1.0	13.8	N, C	N	N	(g)

Remark: N – nepheline, A – albite, C – cristobalite, A – albite, (g) – glassy phase (destruction into a sintered phase).

Table 4.
Phase composition of the fly ash 1-based cements vs. quantity of Na_2O.

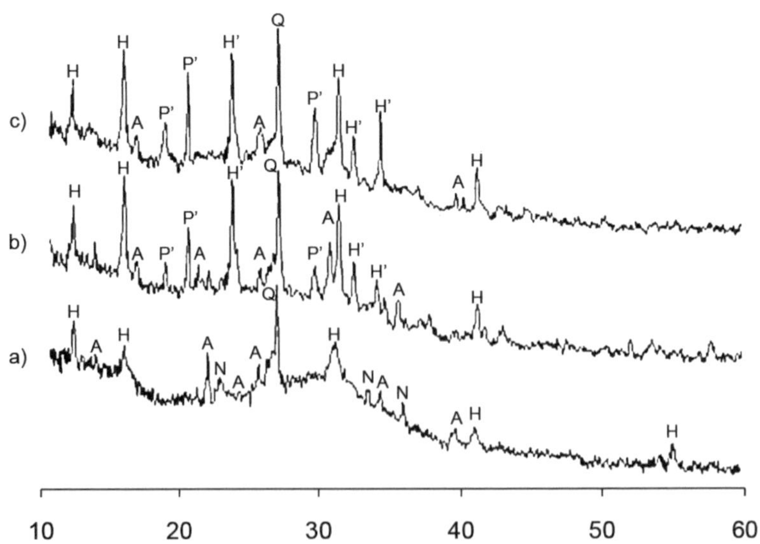

Figure 9.
The XRD scans of the metakaolin-based cements after drying at 80°C (SiO_2/Al_2O_3 = 5 and K_2O/R_2O, respectively: a – 0; b – 0.15; c – 0.3). Q – Quartz; N – Natrolite; A – Zeolite-Na-A; P' – Phillipsite-Na-K; H – Heulandite-Na; H' – Heulandite-K.

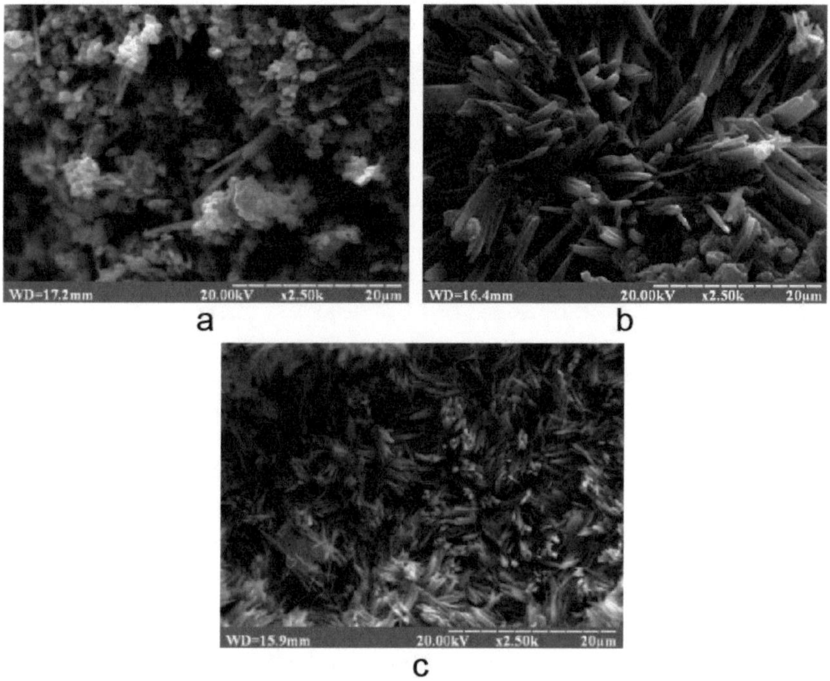

Figure 10.
The SEM images of the microstructure of a cleaved fragment of the metakaolin-based cements after drying at 80°C (SiO_2/Al_2O_3 = 5 and K_2O/R_2O, respectively: a – 0; b – 0.15; c – 0.3)(×2500).

microstructure of the cement matrix displays a large number of sub-crystal phases (**Figure 10**).

The phase composition of the alkali-activated aluminosilicate cements (at K_2O/R_2O = 0) is characterized by the zeolite-like hydration products of the following type: natrolite, zeolite-Na-A and heulandite-Na. When potassium ions are replaced by the sodium ions, other zeolite-like hydration products of the potassium heulandite and sodium-potassium phillipsite types appear as well. The presence of the natrolite phase is not observed, the crystallinity of the hydration products increases.

The cements made using sodium compounds (sodium-based cements) are the most widely used cements among the possible alkali-activated aluminosilicate cements due to the lowest price of the alkaline component. The potassium compounds show in some cases better results than the sodium compounds. In particular, potassium aluminosilicates are more thermally resistant compared to the sodium ones: for example, a temperature of thermal destruction (fusion) of orthoclase ($K_2O{\cdot}Al_2O_3{\cdot}6SiO_2$) is 1170°C (against 1118°C of albite ($Na_2O{\cdot}Al_2O_3{\cdot}6SiO_2$)). A well-known fact is that the binary systems like the sodium/potassium system are significantly less thermally resistant compared to pure systems: eutectic solutions are formed in such systems and these mixed systems fuse at the significantly lower temperatures [37]. The same refers to other binary systems (Na/Ca or K/Ca) and a ternary system (Na/K/Ca). It is important in view of wide application of the Ca-containing compounds as cement modifiers: this technology helps to solve a lot of problems associated with ordinary cements [20] but, being applied for making heat-resistant materials, the increased content of Ca may cause low temperature of sintering and high thermal shrinkage. It is to be noted that this rule is only in power at limited contents of CaO (approx. below 10%), because at the higher calcium contents the system begins revealing other behaviour and, in fact, is not itself the alkali-activated aluminosilicate cement.

Curing temperature, (°C)	Reaction products after curing		
	N-A-S-(H)	K-A-S-(H)	N-K-C-A-S-(H)
400–600	none	—	—
600–900	nepheline	mainly leucite, kaliophilite	—
> 900	albite		leucite, then plagioclase
Description: N – Na_2O; K – K_2O; C – CaO; A – Al_2O_3; S – SiO_2; H – H_2O.			

Table 5.
Phase composition of the fly ash 1-based cements after high temperature curing vs. cation type.

Formation of the phases in case of the fly ash-based cement, depending upon a type of the cation and curing temperature is shown in **Table 5**: the first phases which appear during heating (at approx. 600–900°C) are low-silica feldspathoids (nepheline in case of the sodium-based cements and leucite in case of the potassium-based ones), whereas high-silica feldspars appear at the higher temperatures (900–1200°C). Thus, the earlier collected data on alkaline activation of various clays [4, 28] showed that a phase composition of the synthesized aluminosilicates of alkali metals could also be regulated by changing a temperature of high-temperature curing. Nepheline is a main reaction product of the system composed of a clay activated by sodium compound after curing at 750–900°C, and albite – after curing at the higher temperatures (900–1000°C). When clays are activated by the potassium compounds, leucite is the main reaction product after high temperature curing [28].

Sodium and calcium compounds result in the synthesis of plagioclases (binary sodium-calcium feldspars such as labradorite). Such results were obtained in the potassium (KOH)- and calcium (OPC clinker)-modified cements [20, 21, 27, 29, 30], in which leucite (a potassium feldspathoid) and labradorite (a sodium-calcium feldspar) are the main phases after high- temperature curing (**Figure 11**). This composition, N-K-C-A-S-H, has an amorphous structure after curing at high temperature (800°C). With temperature increase up to 900°C, according to DTA examination (**Figure 12**), this amorphous phase begins to crystallize. After curing at 1000°C, according to XRD examination, leucite and labradorite occur.

After 1200°C, redistribution of shares of the above listed reaction products begins and, in accordance with their thermodynamic stability at this temperature, a quantity of labradorite increases, whereas that of leucite decreases.

Concluding the above, the formation of a microstructure of the alkali-activated aluminosilicate cements may be expressed by the following scheme:

$$\text{zeolite precursor} \rightarrow \text{crystalline zeolite} \rightarrow \text{(amorphization)} \rightarrow$$
$$\rightarrow \text{feldspathoid} \rightarrow \text{feldspar} \rightarrow \text{melting}$$

3.2 Properties of the alkali-activated aluminosilicate cement-based materials

Formation of properties of the materials made using the zeolite-like cement matrices depends upon a cement composition and curing parameters applied to these materials.

The most important characteristics of these materials are mechanical strength (usually evaluated by a residual strength after high-temperature exposure) and stability of volume (evaluated by a thermal shrinkage). Other characteristics that are also important in some cases are resistance against crack formation, thermal resistance and a temperature of deformations under load.

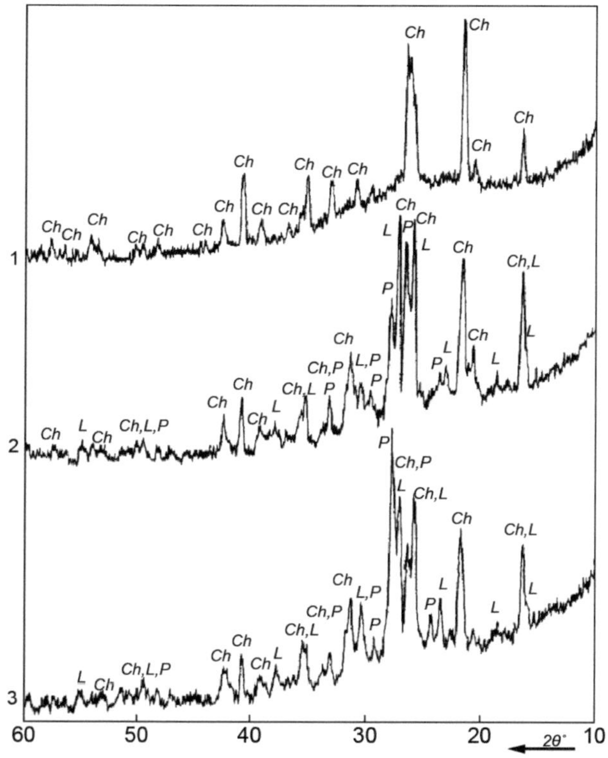

Figure 11.
The XRD scans of the metakaolin-based cement with 10% OPC clinker by mass after curing at 800°C (1), 1000°C (2) and 1200°C (3). (Chamotte (Ch) as a filler)).

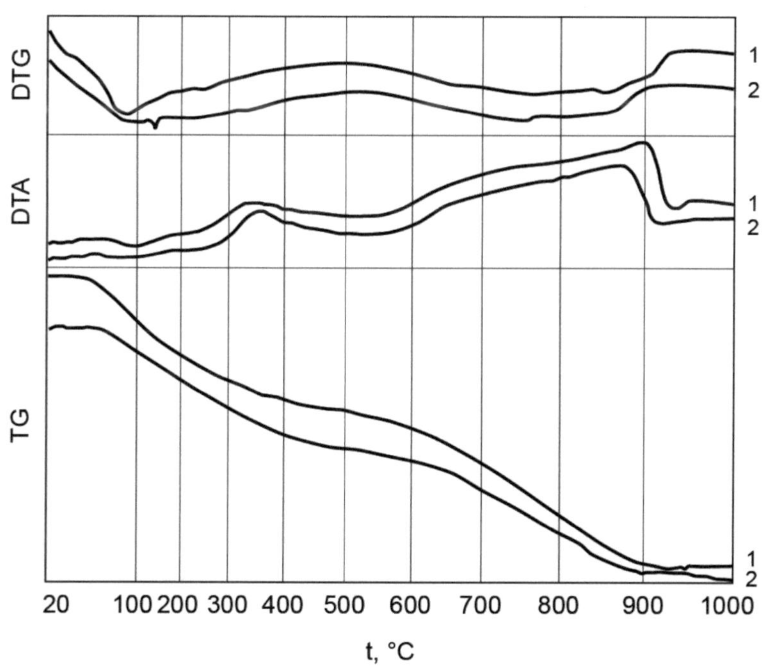

Figure 12.
The TGA, DTG and TG of the hydrated metakaolin-based cements. (Chamotte as a filler): 1 – Without OPC clinker; 2 – With 10% OPC clinker by mass.

Compressive strength and thermal shrinkage change with the temperature increase. When thermally stable zeolite-like phases are the main reaction products of the cement matrices, compressive strength and thermal shrinkage change slowly within a temperature range of 200–800°C. Above 800°C, a liquid phase starts to appear. As a result, the strength increases and the shrinkage also increases too (**Figure 13**). Once the thermal shrinkage reaches the values of 1–2%, the materials lose their functionality because of crack formation, instability of volume, etc., despite still high values of compressive strength. A temperature at which a limit value of thermal shrinkage is reached is a maximum use temperature for a material. The use temperature depends chiefly upon alkalinity and a cation type of the alkali compound. The best heat resistant alkali-activated aluminosilicate cement-based materials may withstand temperatures up to 1000°C).

Heat-resistant materials require a stability of volume changes within a wide temperature range. The shrinkage/expansion processes during exposure to temperatures can be regulated by a choice of an optimal cement composition and appropriate heat-resistant fillers, for example, chamotte [48]. The study of interrelation between a cement composition, curing parameters, phase composition of the reaction products and properties of the cement-based materials was carried out on the compositions: cement: chamotte = 1:1 by mass (**Figure 14**). Also, an interrelation between a relative intensity of the crystalline phase formation and the above properties was studied. In this experiment, a maximum height of the peak in the XRD patterns (fixed for the analcime main peak at 0.343 nm in the XRD pattern of the autoclaved fly ash 1 – based cement with SiO_2/Al_2O_3 = 4) was taken as 100% (**Figure 15**).

The study of interrelation between cement composition, curing conditions, phase composition of the microstructure and properties of the artificial stone was carried out on the compositions (cement:chamotte = 1:1). It was also found that the 1:1 ratio for the composition "fly ash-based cement – chamotte" was optimal for meeting the requirements with regard to workability retention of the mix and

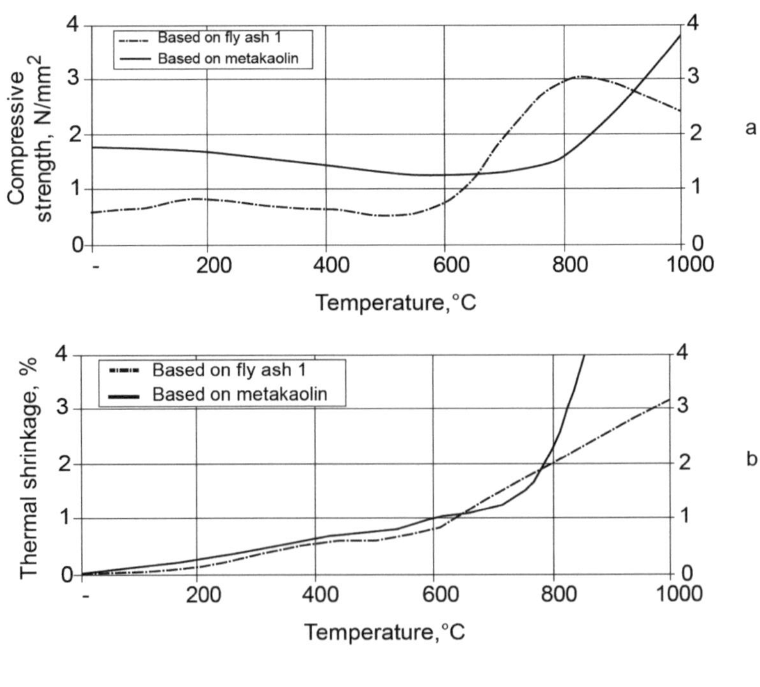

Figure 13.
Compressive strength (a) and thermal shrinkage (b) of the alkali-activated aluminosilicate cement-based materials vs. curing temperature.

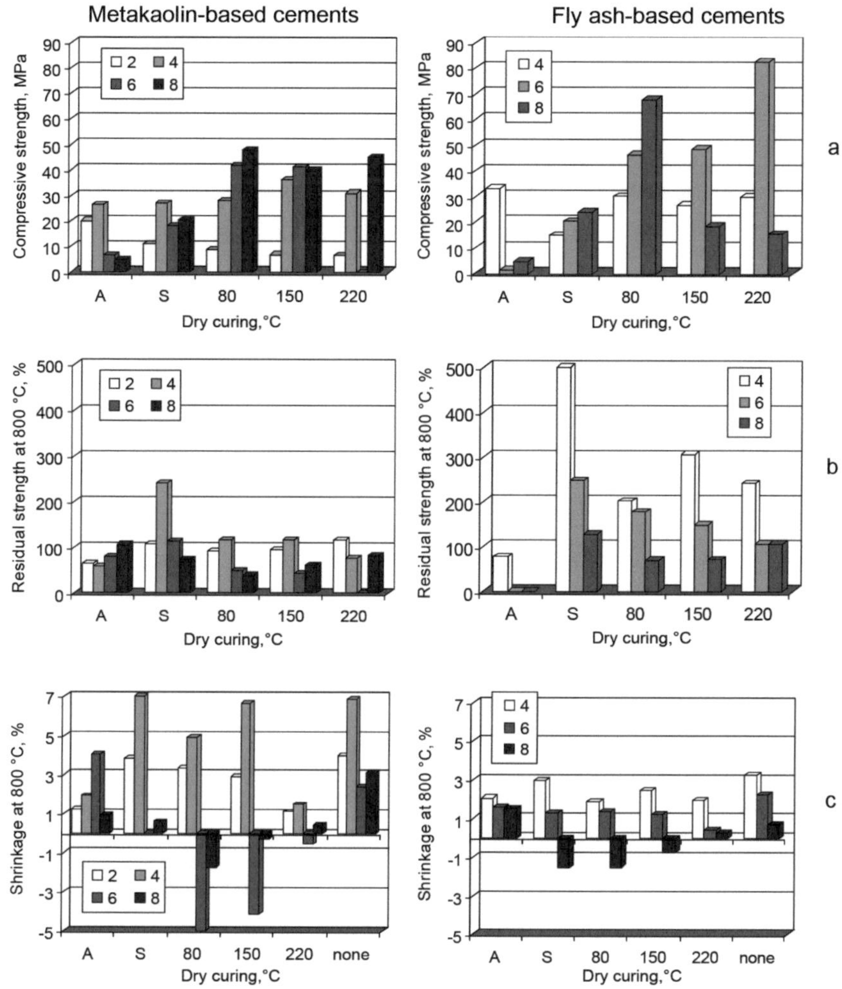

Figure 14.
Compressive strength (a), residual strength (b), and shrinkage (c) after high-temperature curing at 800°C (c) molar ratio Al_2O_3/SiO_2 = 2, 4, 6, 8 of the cements and mode of pre-curing. (A – Autoclave curing at 174°C, S – Steam curing at 80°C, none – Without pre-curing).

compressive strength of the materials both at normal and high temperatures. The higher contents of chamotte the worser are these properties, however, in this case high contents of chamotte help to control shrinkage at high temperatures.

A material made using the alkali-activated aluminosilicate cements with crystalline phases is less inclined to sharp changes in plastic deformations. A crystallization in a ceramic matrix of the compound of the $R_2O{\cdot}Al_2O_3 \times (2{\div}6)SiO_2$ composition, which is analogous to natural nepheline, caliophilite, albite, and orthoclase, is sufficient to produce the materials with properties of ceramics.

Facing materials can be produced using mixed alkali-alkaline earth-activated aluminosilicates of the $mNa_2O{\cdot}(1\text{-}m)\ CaO{\cdot}Al_2O_3{\cdot}nSiO_2$ composition. A conclusion was made that with the increase in the content of the phase of the albite composition ($m \rightarrow 1$) in the reaction products the materials with properties similar to those of facade tiles (water absorption below 10%), and with the increase of the phase of the anorthite composition ($m \rightarrow 0$) – with the properties similar to those of wall ceramics can be produced.

High intensity of crystal formation (higher than 10%) in the compositions after autoclave pre-curing side by side with insufficient degree of crystallization

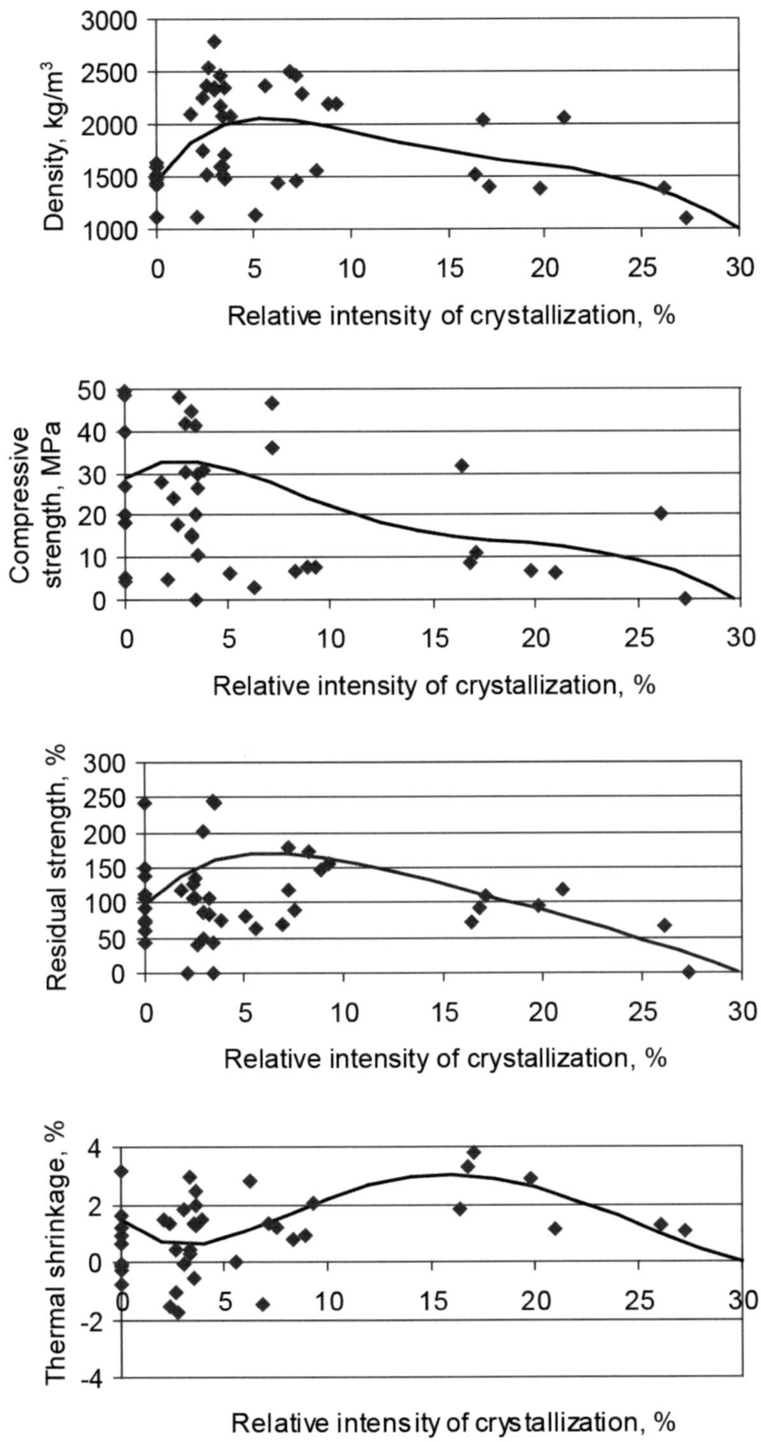

Figure 15.
Relationship between the relative intensity of crystallization and properties of the cement –chamotte compositions: Density, compressive strength after normal temperature curing, residual strength and thermal shrinkage after curing at 800°C.

results in sharp deterioration of the performance properties. In case of low degree of crystallization the structure is characteristic of a weak crystalline framework.

Cement composition		Filler
Aluminosilicate component	**SiO_2/Al_2O_3**	
Metakaolin	4.0	Fly ash 1
Metakaolin	4.0	Fly ash 2
Fly ash 1	4.5	Chamotte
Fly ash 2	5.0	Chamotte

Table 6.
Optimal mixes of the heat resistant composite alkali-activated aluminosilicate cement-based materials.

Taking into consideration the higher reaction ability of metakaolin compared to the fly ash at initial stages of interaction, two systems of heat resistant composite materials were proposed (**Table 6**). These optimized compositions have the initial setting time at 80°C – up to 4 hours, compressive strength – up to 89 MPa, residual strength after high-temperature curing – up to 245% and thermal shrinkage – up to 4.2%. The developed compositions were used for making heat-resistant composite materials for high use temperatures without pre-treatment.

3.3 Experience of application of the alkali-activated aluminosilicate cements

The developed alkali-activated aluminosilicate cement-based materials were used for making a wide range of composite materials [13, 18, 20, 23, 25, 35–37, 44, 47–55] for the use in normal conditions and at high temperatures.

Some examples of production and application of the materials based on the alkali-activated aluminosilicate cements are given below.

3.3.1 Heat insulating materials

- **fibre-reinforced materials (mineral wool, superfine, thin and thickened basalt fibres).** Density of these materials is 90–400 kg/m^3, binder content – 2–30% by mass, heat conductivity at 25°C – 0.044–0.056 W/(m·K), compressibility (at P = 0.002 MPa) – 2–35%, temperature resistance of the ready products – within a temperature range of 600–850°C, depending on a fibre type, contents of sorption moisture and water absorption, of the materials meet the Ukrainian standard-specified requirements.

- **with expanded rocks (perlite and vermiculite).** Inorganic nature of these materials and high temperature resistance determines their high fire resistance and fire safety. These materials have density – 150–300 kg/m^3, compressive strength –0.26–1.0 MPa, flexural strength – 0.15–0.3 MPa, heat conductivity – 0.062–0.09 W/(m·K).

- **with mullite-silica fibres** have mean density 310–420 kg/m^3, flexural strength – up to 0.4 MPa, the lower heat conductivity at 600°C (0.18–0.22 W/(m·K)) compared to commercial products, the lower loss on ignition (0.01–1%), the increased content of Al_2O_3 + SiO_2 (97–98%) and possess the higher temperature resistance (1300°C).

- **gas concretes** exhibit compressive strength – 2–5 MPa, density – 400–550 kg/m^3, much higher temperature resistance – up to 800°C compared to the traditional gas cement concretes. These materials can be recommended for the use in not only in construction, but for the production of insulation of industrial equipment working at high temperatures.

3.3.2 Constructive materials

- **with basalt rovings** have density 850–1000 kg/m^3, surface density – 2.6–9.85 kg/m^2, binder content –42–49% by mass, organic matters content of the final product 0–10% by mass, flexural strength – 28–30 MPa, maximum use temperature – 250–500°C. They have equal or considerably higher maximum use temperature (up to 500°C).

- **with glass- and basalt non-woven materials and fabrics** have density 900–1800 kg/m^3, binder content – 21–38% by mass, contents of organic matters of the final product – 5–12% by mass, flexural strength – 94–130 MPa, maximum use temperature – 300–400°C, maximum temperature of short-time exposition to high temperatures – 400–500°C.

- **with mullite-siliceous fibres** have density 1600–1700 kg/m^3, binder content – up to 25% by mass, flexural strength – up to 11 MPa, maximum use temperature – 1000°C.

- **with carbon fabrics** have density 2200 kg/m^3, binder content – 50% by mass, organic matters content of the final product – 11% by mass, flexural strength – up to 110 MPa, maximum use temperature – 300°C.

- **with expanded rocks** have density 640–1600 kg/m^3, binder content – up to 25% by mass, compressive strength 1.0–15.0 MPa, flexural strength 2.8–10.0 MPa. These materials possess high heat resistance (residual strength after firing at 800°C is up to 225%).

- **with organic fabrics (jute)** have density 1000–1250 kg/m^3, flexural strength 30–45 MPa, Young's modulus – 2–14 $\times$ 10^9 Pa, temperature resistance – up to 105°C.

An example of industrial application of the constructive alkali-activated aluminosilicate cement-based materials is a body of a fire fighting bomb. The use of the developed constructive alkali-activated aluminosilicate cement-based materials and basalt fabrics allowed to replace a metallic body of the aerosol forming bomb in fire-fighting, which is heated in the process of blasting and is a source of secondary ignition. Field testing of the bomb showed its efficiency in fire-fighting. The material of the bomb body withstands mechanical and thermic loads occurring in the process of operation and is not a source of secondary ignition.

3.3.3 Abrasion resistant composite materials

The abrasion resistant composites with high thermo-mechanical characteristics and instruments from them were developed through modification of the alkali-activated aluminosilicate cement composition by a silicon carbide. The material of the instruments that were produced commercially in Ukraine had the following properties: compressive strength – 47–50 MPa, flexural strength – 23–25 MPa, breaking strength – 5–7 MPa, thermo-resistance – 25 heat cycles "800 °C – water".

3.3.4 Glues for wood-based materials

The use of the alkali-activated aluminosilicate cements as a binding agent or adhesive in the production of wood-based materials permits not only to replace toxic organic resins, but to enhance considerably fire resistance.

One more example is the use of the alkali-activated aluminosilicate cements in the production of wood-based materials (chip wood aggregate concretes and wood particle boards).

The wood particle boards have densities from 500 to 1500 kg/m^3, flexural strength – 8–20 MPa, swelling – less than 1% for 24 hours, the cement content in them – up to 40% by mass of the ready product.

3.3.5 Binders for metallurgical industry

High thermal resistance of the alkali-activated aluminosilicate cement-based materials allowed to develop a range of products for the needs of metallurgical industry: pipes for feeding metal melt, sieves for melt filtration, heat insulating feeders, etc. The use of these materials allows for to replace high-cost ceramics. Above all, the developed compositions for moulding and core sands to be used in the foundry permit to replace highly toxic organic adhesives. The developed materials are characteristic of high green strength (up to 0.25 MPa) and strength in a dry state – up to 7.5 MPa, required minimal strength of a moulding sand after metal is fed, high values of sand reclamation (80–90%), thus meeting the requirements applied to moulding sands.

Intumescent fireproofing and expanded heat insulating coatings based on the alkali-activated aluminosilicate cements have coefficient of volume increase – 15–23, compressive strength – up to 4.6 MPa, water absorption – below 10% by mass, adhesion to metal – 0.78–0.84 MPa and ceramics – 4.8–5.0 MPa. The expanded coatings are characteristic of coefficient of heat conductivity (λ)–0.041–0.065 W/m·°C. Fireproofing coatings for wood, wood-based materials and fibre-reinforced boards are characteristic of resistance in an open flame (800–1000°C) for 10–20 minutes. The tests and introduction at the enterprises of Ukraine showed their efficiency in replacing traditional compositions made using organic components.

Corrosion resistant coatings made with using the alkali-activated aluminosilicate cements have compressive strength after hardening within a temperature range of 20–150°C – 50–110 MPa, changes in mass and volume after boiling in 2% milk acid solution – 1.5–2.0%, whereas in 1–3% solutions of sulphuric acid – 2–2.5% with a coefficient of resistance 0.85–0.97 in organic mediums and 0.7–0.78 in sulphuric acid. The use of autoclave treatment allows to produce a material, which is able to withstand exposure 5–10% concentrations of these chemicals, coefficient of resistance after attacks of organic chemicals being 0.98–1.00, of sulphuric acid – 0.8–1.0 and volume changes – 2.0–2.5%. Adhesion between the coating and concrete – 3–8 MPa. The composites with the enhanced acid resistance were used in a two-layered coating of the external lining of a sewage waters disposal canal. The coatings can be recommended for service in the solutions of sulphuric acid (with concentrations below 5%), pickling solutions and solutions of heavy metal salts.

3.3.6 Glues for repairing and restoration

A variety of the alkali-activated aluminosilicate cement-based materials for repairing and restoration needs was developed: primers and putties etc.

Materials to be glued together	Ratio of summary areas in non-uniform breaking, %		
	Cohesion failure (in adhesive), no more than	Adhesion failure (Interaction zone "material-adhesive"), more than	Failure in fibre, not less than
Heat insulating board – paperboard	5	5	90
Paperboard – steel sheet	20	30	50
Heat insulating board – steel sheet	20	30	50
Heat insulating board – aluminium sheet	20	30	50

Table 7.
Properties of the adhesive joints made using the alkali-activated aluminosilicate cement.

3.3.7 Inorganic adhesives

One of the most effective application of the alkali-activated aluminosilicate cements is to use them as inorganic binders for anchoring application, for bonding together various materials: mineral fabrics, fibre-reinforced heat insulating articles and paper, concrete, brick and ceramics, wood with concrete and gypsum, for metal, expanded polystyrene, plastics etc. The adhesive joints exhibit strength in non-uniform pulling-out – 3–15 MPa, high freeze–thaw resistance (more than 100 cycles) and water resistance. Examples of the use of these adhesives are given in **Table** 7.

3.3.8 Acid resistant ceramic materials

Basing on the mixed albite-orthoclase compositions of the reaction products that are used as binders and silica sand as filler, the acid resistant ceramic materials with the properties of traditional acid resistant materials based on clays and those very

Characteristics	Materials based on alkali-activated aluminosilicate binders	Cast basalt	Ceramics based on clays
Water absorption, %	0.2–5.0	0.05	2–10
Volume mass, kg/m^3	2200–2400	3000	2000–2170
Strength, MPa			
compressive	110–200	250–300	30–90
flexural	25–40	50–65	10–20
tensile	12–20	30	6–10
Acid resistance, %	98.0–99.6	99.0–99.6	95.0–98.0
Thermal resistance, a number of heating-sharp cooling cycles	5–30	3–3	2–20
Wear abrasion, kg/m^2	0.30–0.50	0.15–0.25	0.60

Table 8.
Comparative physico-mechanical characteristics of the acid resistant ceramic materials.

close to cast basalt (**Table 8**) can be produced. A curing temperature is between 1150 and 1200°C.

Testing by boiling of the specimens in H_2SO_4, HCl, and HNO_3 for 700 hr. showed that with the longer duration the rate of degradation of the material tended to decrease, on the contrary to the acid resistant ceramics based on clays, the process of degradation of which in the same conditions became more intensive.

Water absorption, mechanical strength and acid resistance of the materials after storage for 12 months in H_2SO_4 -solutions (98, 60, and 30% concentration), in HCl-solutions (30, 20, and 10% concentrations) and in HNO_3-solutions (60, 30, and 10% concentrations) suggested to conclude that corrosion of the acid resistant materials based on the alkali-activated aluminosilicate cements tended to decrease. Long-term testing showed high resistance of these materials not only in mineral acids, but in chlor-containing air media.

4. Conclusions

The zeolite-like analogies to natural aluminosilicate minerals may be synthesized in the cement matrices of the Na(K)-Al_2O_3-SiO_2-H_2O system. Hydrous zeolites are formed in this system after 80 °C, after 600°C (within a temperature interval of 600–1200°C) they transform into anhydrous zeolites. The structure formation and properties of the alkali-activated aluminosilicate cement-based materials at low temperatures may be regulated by changing a cement composition, curing conditions, type and concentration of the alkaline activator solution as well as the solution to solid ratio. A directed regulation of the low-temperature structure formation process is a key instrument allowing to obtain a wide range of special materials using various types of the alkali-activated aluminosilicate cement and curing conditions. At high temperatures, the only significant factors affecting a phase composition and properties are an initial cement composition and a curing temperature. Regulation of the high-temperature structure formation is important in order to develop heat- and fire- as well as acid resistant materials. High durability of the alkali activated cement matrix can be attributed to a synthesis of the zeolite-like analogies to natural minerals. Excellent performance properties and high durability of the alkali-activated aluminosilicate cement-based materials for high temperature uses are provided due to smooth dehydration and subsequent re-crystallisation of the hydration products into stable anhydrous aluminosilicates of alkali metals.

Author details

Pavel Krivenko* and Volodymyr Kyrychok
Scientific Research Institute for Binders and Materials, Kyiv National University of Construction and Architecture, Kyiv, Ukraine

*Address all correspondence to: v.kyrychok@gmail.com

References

[1] Malinowski R. Concretes and mortars in ancient aqueducts. Concrete International. 1979; (1)1: 66–76.

[2] Davidovits J. Design and Construction. Concrete International. 1987; (9)12: 23.

[3] Krivenko P. Why alkali activation – 60 years of the theory and practice of alkali- activated materials. Ceram. Sci. Technol. 2017; 8(3): 3203–333. https://doi.org/10.441/JCST2017-00042

[4] Glukhovsky V. Durability of Concrete. In Proceedings of the Intern. Seminar. Götheborg; 1989. p. 53–62.

[5] Breck D. Zeolite Molecular Sieves. J. Wiley & Sons: New York; 1974.

[6] Barrer R. Hydrothermal Chemistry of Zeolites. Academic Press: London; 1982.

[7] Dyer A. An Introduction to Zeolite Molecular Sieves. J. Wiley & Sons: New York; 1988.

[8] Davidovits J. Geopolymer chemistry and properties. In Proceedings of the 1st European Conf. on Soft Mineralurgy "Geopolymer'88"; Saint-Quentin; 1988. p. 25–48.

[9] Krivenko P. Alkaline cements: terminology, classification, aspects of durability. In Proceedings of the Tenth International Congress on the Chemistry of Cement; Götheborg; 1997. p. 4iv046-4iv050.

[10] Sukhanevich M., Guzii S. The effect of technological factors on properties of alkali aluminosilicate systems used for preparation of fireproof coatings. Refractories and Industrial Ceramics. 2004; 45(3): 217–219. https://doi.org/10.1023/B:REFR.0000036733.85631.f1

[11] Krivenko P., Pushkareva Y., Sukhanevich M., Guziy S. Fireproof coatings on the basis of alkaline aluminium silicate systems. Ceramic Engineering and Science Proceedings. 2009; 29(10): 129–142.

[12] Krivenko P., Guziy S. Aluminosilicate coatings with enhanced heat- and corrosion resistance. Applied Clay Science. 2009; 73(1): 65–70. doi: 10.1016/j.clay.2012.10.010

[13] Kovalchuk G., Krivenko P. Producing fire-and heat-resistant geopolymers. In: Geopolymers: Structures, Processing, Properties and Industrial Applications. Woodhead Publishing; 2009. p. 227–266

[14] Kravchenko A., Guzii S. Determining the fire resistance properties of timber, protected by geocement-based coatings. Eastern European Journal of Enterprise Technologies. 2015; 1(5): 38–41. https://doi.org/10.15587/1729-4061.2015.36843

[15] Kryvenko P., Tsapko J., Guzii S., Kravchenko A. Determination of the effect of fillers on the intumescent ability of the organic-inorganic coatings of building constructions. Eastern European Journal of Enterprise Technologies. 2016; 5(10–83): 26–31. https://doi.org/10.15587/1729-4061.2016.79869

[16] Tsapko Y., Kyrycyok V., Tsapko A., Bondarenko O., Guzii S. Increase of fire resistance of coating wood with adding mineral fillers. MATEC Web of Conferences. 2018; 230: 20–34. doi: 10.1051/matecconf/201823002034

[17] Xu H., Van Deventer J. In Proceedings of the Second International Conf. on Geopolymers "Geopolymer'99"; Saint-Quentin; 1999. p. 43–64.

[18] Krivenko P., Skurchinskaya J. In Proceedings of the Intern. Conf. on the

Utilization of Fly Ash and other Coal Combustion By-Products. Shanghai. 1991. p. 64–1 – 64-7.

[19] Duan P, Yan C, Zhou W., Luo W. Thermal Behaviour of Portland Cement and Fly Ash-Metakaolin-Based Geopolymer Cement Pastes. Arab. J. Sci. Eng. 2015; 40: 2261–69. DOI 10.1007/s13369-015-1748-0

[20] Pangdaeng S., Phoo-ngernkham T., Sata V., Chindaprasirt P. Influence of curing conditions on properties of high calcium fly ash geopolymer containing Portland cement as additive. Materials& Design. 2014; 53: 269–274. doi:10.1016/j.matdes.2013.07.018

[21] Bernal S, Rodríguez D, De Gutiérrez R, Gordillo M, Provis J. Mechanical and thermal characterisation of geopolymers based on silicate-activated metakaolin/slag blends. Mater. Sci. 2011; 46: 5477.

[22] Duxson P, Provis J, Lukey G, Mallicoat S, Kriven W, Van Deventer J. Understanding the relationship between geopolymer composition, microstructure and mechanical properties. Colloids and Surfaces. Physicochem. 2005; 269: 47–58. doi: 10.1016/j.colsurfa.2005.06.060

[23] Barbosa V F, MacKenzie K J. Thermal behaviour of inorganic geopolymers and composites derived from sodium polysialate. Mater. Res. Bulletin. 2003; 38: 319–31. doi:10.1016/S0025-5408(02)01022-X

[24] Saavedra W.G.V., De Gutierrez R. M. Performers of geopolymer concrete composed of fly ash after exposure to elevated temperatures. Construction and Building materials. 2017; 154: 229–235. https://doi:10.1016/j.conbuildmat.2017.07.208

[25] Lee N., Koh K., An G., Ryu G. Influence of binder composition on the gel structure in alkali activated fly ash/slag pastes exposed to elevated temperatures. Ceramics International. 2017; (43)2: 2471–2480. doi:10.1016/j.ceramint.2016.11.042

[26] Rivera O., Long W., Weissjr C., Moser R., Williams B., Torres-Cacel K., Gore E., Allison P. Effect of elevated temperature on alkali-activated geopolymeric binders compared to Portland cement-based binders. Cement and Concrete Research. 2016; 90: 43–51. https://doi:10.1016/j.cemconres.2016.09.013

[27] Wang W-Ch, Wang H-Yu, Lo M-h. The engineering properties of alkali-activated slag pastes exposed to high temperatures. Construction and Building materials. 2014; 68: 409–415. https://doi:10.1016/j.conbuildmat.2014.06.016

[28] Hlaváček P, Šmilauer V, Škvára F, Kopecký L, Šulc R. Inorganic foams made from alkali-activated fly ash: Mechanical, chemical and physical properties. Eur. Ceram. Soc. 2015; 357: 03–09. doi:10.1016/j.jeurceramsoc.2014.08.024

[29] Zhang H, Kodur V, Qi S, Cao L, Wu B. Development of metakaolin–fly ash based geopolymers for fire resistance applications. Constr. Build. Mater. 2014; 55: 38–45. doi:10.1016/j.conbuildmat.2014.01.040

[30] Rovnaník P, Bayer P, Rovnaníková P. Characterization of alkali activated slag paste after exposure to high temperatures. Constr. Build. Mater. 2013; 47: 1479–87. DOI:10.1016/j.conbuildmat.2013.06.070

[31] Bakharev T. Thermal behaviour of geopolymers prepared using class F fly ash and elevated temperature curing. Cem. Concr. Res. 2006; 36: 1134–1147. doi:10.1016/j.cemconres.2006.03.022

[32] Sitarz M., Hager I., Kochanek J. Effect of High Temperature on

Mechanical Properties of Geopolymer Mortar. In: Proceedings of the MATBUD'2018 – 8th Scientific-Technical Conference on Material Problems in Civil Engineering MATEC; 15 June 2018; MATBUD; 2018. doi: 10.1051/matecconf/201816306004

[33] Su H., Xu J., Ren W. Mechanical properties of geopolymer concrete exposed to dynamic compression under elevated temperatures. Ceramics International; 2017; (43)2: 2471–2480. doi:10.1016/j.ceramint.2015.11.055

[34] Zulkifly K., Young H.C., Abdullah M.M.A B., Ming L.Y., Panias D., Sakkas K. Review of Geopolymer Behaviour in Thermal Environment. In: Proceedings of the Int. Conf. on Innovative Research–ICIR EUROINVENT 2017; Materials and Engineering Vol. 209 (2017) Doi: 10.1088/1757-899X/209/1/01/2085

[35] Sotiriadis K., Guzii S., Kumpová I., Mácová P., Viani A. The effect of firing temperature on the composition and microstructure of a geocement-based binder of sodium water-glass. Solid State Phenomena. 2017; 267: 58–62. doi: 10.4028/www.scientific.net/SSP.267.58

[36] Sotiriadis K., Guzii S., Macova P., Viani A., Dvorak K., Drdacky M. Thermal Behaviour of an Intumescent Alkaline Aluminosilicate Composite Materials for Fire Protection of Structural Elements. Mater. Civ. Eng. 2019; 31(6): 04019058. Doi:10.1061/(ASCE)MT.1943-5533.0002702

[37] Krivenko P., Guzii S., Bodnarova L., Valek J., Hela R., Zach J. Effect of thickness of the intumescent alkali aluminosilicate coating on temperature distribution in reinforced concrete. Journal of Building Engineering. 2016; 8: 14–19. doi:10.1016/j.jobe.2016.09.003

[38] Duxson P., Provis J., Lukey G., Separovic F., van Deventer J. 29Si NMR study of structural ordering in aluminosilicate geopolymer gels. Langmuir. 2005; 21(7): 3028–3036.

[39] Van Deventer J. S. J., Provis J. L., Duxson P., Lukey, G. C. Reaction mechanisms in the geopolymeric conversion of inorganic waste to useful products. Journal of hazardous materials. 2007; (139)3: 506–513.

[40] Davidovits J. Geopolymeric Reaction in Archeological Cements and in Modern Blended Cements. In: Proceedings of the Geopolymer '88; 1988 Jun 1–3; Compiègne, France. p. 93–105.

[41] Duxson P., Lukey G., Separovic F., Van Deventer J. Effect of alkali cations on aluminum incorporation in geopolymeric gels. Industrial & Engineering Chemistry Research. 2005; 44(4): 832–839.

[42] Davidovits J. Geopolymers: inorganic polymeric new materials. Journal of Thermal Analysis and calorimetry. 1991; 37(8): 1633–1656.

[43] Swaddle T., Salerno J., Tregloan P. Aqueous aluminates, silicates, and aluminosilicates. Chemical Society Reviews. 1994; 23(5): 319–325.

[44] Petranek V., Krivenko P., Petropavlovskiy O., Guzii S. Perlite concrete based on alkali activated cements. Advanced Materials Research. 2014; 897: 280–283. doi:10.4028/www.scientific.net/AMR.897.280

[45] Palomo Á., Alonso S., Fernandez-Jiménez A., Sobrados I., Sanz J. Alkaline activation of fly ashes: NMR study of the reaction products. Journal of the American Ceramic Society. 2004; 87(6): 1141–1145.

[46] Fernández-Jiménez A., Palomo A., Sobrados I., Sanz J. The role played by the reactive alumina content in the alkaline activation of fly ashes. Microporous and Mesoporous materials. 2006; 91(1–3): 111–119.

[47] Turker H., Alcikani M., Durmus I., Ozbay E., Erdemir M. Microstructural alteration of alkali activated slag mortars depend on exposed high temperature level. Construction and Building materials; 2014: 409–415. https://doi: 10.1016/j.conbuildmat.2015.12.070

[48] Rovnaník P., Šafránková K. Thermal Behaviour of Metakaolin/Fly Ash Geopolymers with Chamotte Aggregate. Materials. 2016; 9: 535. doi: 10.3390/ma9070535 www.mdpi.com/journal/materials

[49] Duxson P, Lukey G, van Deventer J. Physical evolution of Na-geopolymer derived from metakaolin up to 1000 C. Mater. Sci. 2007; 42: 3044–54.

[50] Duxson P, Lukey G, Deventer J. Evolution of gel structure during thermal processing of Na-geopolymer gels Langmuir. 2006; 22: 8750–8757. DOI:10.1021/la0604026

[51] Petránek V., Guziy S., Sotiriadis K., Nevřivová L. Study on the properties of geocement based thermal insulating materials for high temperature technical appliances. Advanced Materials Research. 2013; 734–737: 2356–2359. doi: 10.4028/www.scientific.net/AMR.734-737.2356

[52] Petránek V., Guzii S., Kryvenko P., Sotiriadis K., Kravchenko A. New thermal insulating material based on geocement. Advanced Materials Research. 2014; 838–841: 183–187. doi: 10.4028/www.scientific.net/AMR.838-841.183

[53] Petránek V., Nevrivova L., Zezulova D., Guziy S. Thermal Insulating Materials for Energy Storage Application. Advanced Materials Research. 2014; 911: 30–35. doi:10.4028/www.scientific.net/AMR.911.30

[54] Petránek V., Guzii S., Kryvenko P., Sotiriadis K., Maňák J. Use of thermal insulating perlite composite materials based on geocement to protect technological equipment. Advanced Materials Research. 2014; 860–863: 1342–1345. doi:10.4028/www.scientific.net/AMR.860-863.1342

[55] Krivenko P., Mokhort M., Petropavlovskii O. Industrial Uses of Geocement-based Materials in Construction and Other Industries. In the Proceedings of the Int. Conf. “Geopolymer” (2002).

Chapter 7

Zeolites as Scaffolds for Metal Nanoclusters

Eduard Fron

Abstract

This chapter critically reviews the studies related to structural and photophysical properties of metal clusters within zeolites matrices and summarizes the progress made in understanding the host-guest interactions. The goal is to provide useful insight into the nature of such interactions and experiments used in identifying the excited state dynamics and the reaction mechanisms leading to the emitting species. Especially interesting are the combined experimental and computational approaches used to elucidate the structures and electronic transition of clusters inside the cavity. Although a number of excellent research articles have been published in the last years they only cover rather specific areas like organic photochemistry, confinement, charge transfer, theoretical modeling or photostimulated luminescence.

Keywords: photphysics, structure, time-resolved spectroscopy, absorption, emission

1. Introduction

Nobel metal clusters confined in restricted environment of zeolite possess remarkable absorption and emission properties, large Stokes shifts and, with a few exceptions, exceptionally high external quantum efficiencies (EQE's) which are of paramount importance in various processes and applications [1–10]. In the last years, a converging view is that the origin of their optical properties resides in their molecular-like characteristics as a result of a strong quantum confinement leading to discrete energy levels. However, these intriguing effects appear to depend not only on confinement but also on size, structure, and hydration level, charge state of the cluster and host-guest interactions. Electrostatic interactions between zeolite cavity and confined metal nanoparticles govern the photophysical properties of these materials. Metal clusters self-assembled in the well-defined cavities of aluminosilicate crystalline framework of zeolites possess the most fascinating optical properties because of the complex interaction that the clusters develop with the zeolite framework. This is probably the reason that fundamental research on such materials has been less attractive, perhaps because of the difficulties encountered in rigorous determination of the exact nature of these interactions. A number of questions are frequently asked. What is the nature of the electronic transitions and especially of the long-lived emitting species? Is there a charge transfer between zeolite framework and metal cluster involved in the excited state dynamics? Is the luminescence originating from recombination of electrons trapped or simply from species which are not strongly coupled to the zeolite structure? Is an intersystem crossing occurring upon excitation which

IntechOpen

indicates a forbidden transition/relaxation and thus the long lifetime of the luminescent electronic state? What is the physics that determines which state decays radiatively and can we map the excited state dynamics?

This chapter critically reviews the studies related to the structural and photophysical properties of metal clusters within zeolites matrices and summarizes the progress made in understanding the host-guest interactions. The goal is to provide useful insight into the nature of such interactions, methods and experiments used in identifying the excited state dynamics and the reaction mechanisms leading to the emitting species. Although a number of excellent research articles have been published in the last years they only cover rather specific areas like organic photochemistry, confinement, charge transfer, theoretical modeling or photostimulated luminescence [11–15]. The chapter is organized in three sections relevant to the interplay cluster-framework and seen from a mechanistic point of view that is further supported by various theoretical and experimental based studies like DFT, diffraction or time-resolved luminescence spectroscopy. The first part presents in short the structure and chemical properties of zeolites which is then followed by the progress in understanding the formation and structure of the metal clusters stabilized in the zeolite cavities, pores and channels. The last part sheds light onto the electronic properties and the origin of intense luminescence and how these depend on the interplay between cluster and framework. Especially interesting are the combined experimental and computational approaches used to elucidate the structures and electronic transition of clusters inside the cavity. Particular emphasis is then placed on various debated mechanisms as models to address the quantized electronic interaction which can lead to new insights into their optical, luminescence, crystal habit, metal-core, ligand-shell, and environmental properties [16].

2. Structure and chemical properties of zeolites

As one of the most important materials used in catalysis, adsorption or ion exchange, zeolites have triggered the interest of scientists because of their structural elements, such as cavities, pores and channels as well as their catalytic properties [10, 17–22]. The molecular-sized open framework forming a periodic array of void spaces, enables confinement of guest particles or molecules and opened interesting research areas like metal-to-insulator transition, charge transfer, solvation and production of trapped electrons [12, 23]. Since their discovery in 1756 the molecular sieve properties are being increasingly used in industry in applications as chemical sensors, medical monitoring or air separation, to name a few [11].

The microporous crystalline aluminosilicate consist of 3-dimenisonal Si-O/Al-O bond tetrahedral framework of nearly spherical cages connected through subnanometric windows with alkali or alkaline earth metals as counterions (**Figure 1**) [24–27]. Basically, they consists of a structure of $[SiO_4]^{4-}$ and $[AlO_4]^{5-}$ tetrahedra connected via their oxygens. They have the general formula: $(M^+)_x\ [(AlO_2^-)_x (SiO_2)_y] \cdot mH_2O$ where M^+ indicates the alkali metal cation. Depending on how they are connected, the tetrahedra form a three-dimensional framework with pores, channels and cavities [11].

The framework is negatively charged in the zeolites containing aluminum and this is due to unbalanced charge of the $[SiO4]^{4-}$ and [AlO4] [5] primary building units (PBUs) which needs to be compensated by the cations. The cations and, eventually, water molecules are distributed inside the cavities which, when aligned, become channels. The presence of H_2O molecules inside the cavities is the reason why zeolites can be hydrated/dehydrated by changing the sample temperature. The PBUs combine by sharing oxygens with adjacent tetrahedra to form a spatial

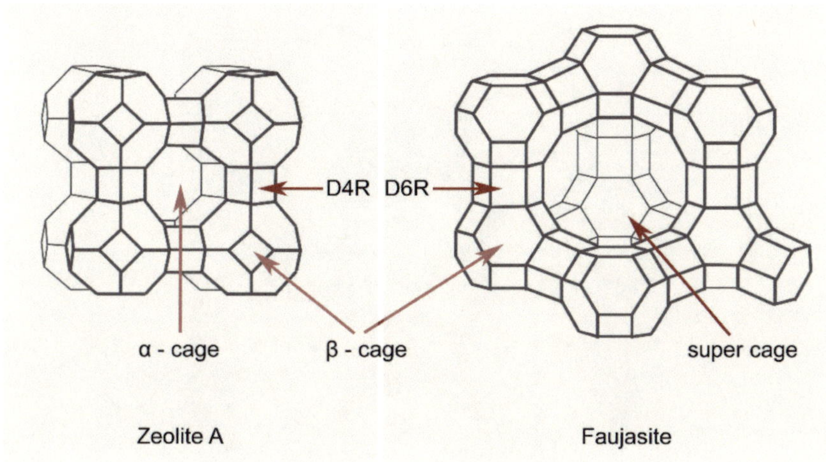

Figure 1.
Framework structure of sodalite-based zeolites.

arrangement of simple geometric forms named the secondary building units (SBU). The crystalline structure is a net product of the special arrangement of SBUs resulting in a large variation in type and morphology of different species of zeolites. The type of zeolites mostly employed in stabilization of metal clusters are A (LTA), FAU (faujasite), and ZSM-5. However, other zeolite morphologies exist while their composition is not limited to the aluminosilicate. For the purpose of this chapter we will limit the description of morphologies to the ones mentioned above.

The structure of zeolite A (LTA) is rather simple and results from connecting the same building unit (β-cage) through a pair of D4R rings (**Figure 1**). Such an arrangement generates a structure with three type of cavities: a D4R square cuboid, an α-cage and of course, the β-cage (**Figure 2**). The largest void present is the α-cage with a diameter of 11 Å and a window of 4.1 Å limiting the size of the potential guests.

Faujasite zeolites exists in two structures X and Y, both constituted from β-cages building blocks. The diameter of the β-cage is 6.6 Å while the window is 2.1 Å and connects to the frame via double D6R. This small window size prohibits the molecular oxygen entering the cage while this remains accessible to water molecules. The difference between X and Y frameworks consists simply in a different Si/Al ratio which is between 1.0 and 1.5 for zeolite X and 1.5 and 3 for zeolite Y. Interestingly, such assembly of β-cages gives rise to a quasi-spherical super cage (**Figure 1**) with a diameter of almost twice the β-cage 13 Å and a window size of 7.4 Å. The mobile counterions needed to compensate the framework charge are distributed inside the β-cage, on the hexagonal faces, at the interface of the supercage 4-ring and D6R unit.

In contrast to LTA and Faujasite topologies, ZSM-5 uses a 5 ring as building unit resulting in a zeolite framework with two type of channels having diameters of 5-6 Å and lengths up to 500 μm. L-type zeolite have a unidimensional channel system as in the mordenite, formed by 6- and 4-membered rings (**Figure 3**).

It is worth mentioning that the electronic structure of the material is in fact a superposition of that of the framework, co-cations, solvent molecules and, eventually, the guest molecules of clusters. The optical bandgap can be as high as 7 eV [2].

The concept of Zeolites as 'solvent' has been introduced by Hashimoto [11] to indicate that the pores and cavities can be used to "dissolve" guest entities. The combination of negatively charged framework and cations possessing a large degree

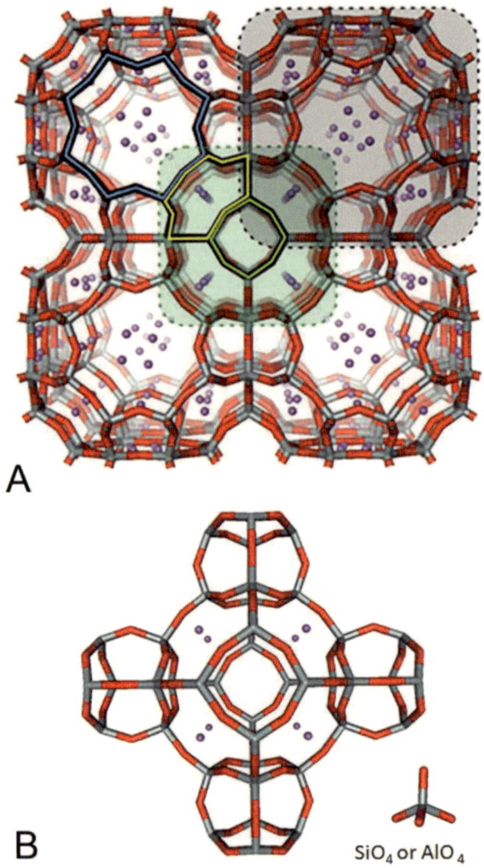

Figure 2.
Schematic representation of (A) LTA unit cell displaying the sodalite (green dashed area) and super cage (gray dashed area); the eight ring (8R), six ring (6R), and double four rings (D4R) are highlighted in blue, yellow and green respectively, (B) isolated sodalite cage with D4R connectors.

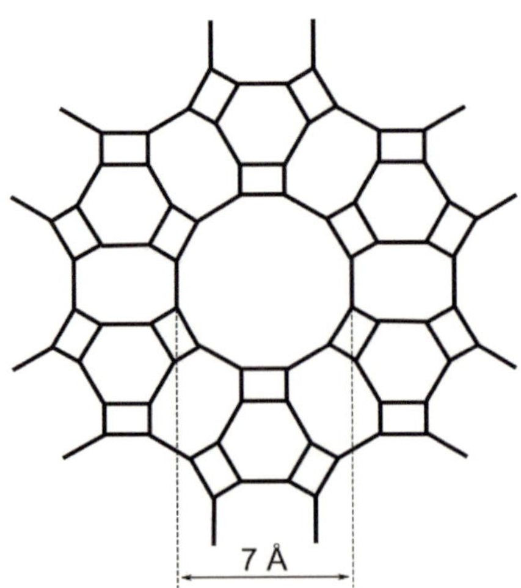

Figure 3.
The block unit of zeolite L.

of freedom produce an electrostatic field in which the guest particles are dispersed. Locally, the electric field strength is believed to be extremely high as its intensity hinges on both cation size and Si/Al ratio [28]. In this light, a large number of molecules have been encapsulated into the zeolite cavities to probe the local electrostatic field and the effects induced by the zeolite framework [29–31].

Exceptional photophysical and photochemical properties are observed when the zeolite crystal dimension is reduced from micrometer size to extremely small nanometer particles with a distribution centered at 10–15 nm [1]. While the micropore volume remains comparable to their corresponding micrometer-sized crystals, the resulted high external surface area, 180 $m^2 g^{-1}$, opens a number of opportunities for processes taking place specifically on this part of the zeolite. The presence of silanes changes the Si/Al ratio and the surface charge of the nanosized zeolites and also the crystal size [8].

3. Formation and structure of the metal clusters in zeolite framework

3.1 Progress

Already in the years 1970s, the high temperature hydrogen reduction of Ag ion to metallic silver in zeolite Y was known but it was only in 1978 when Kellerman and Texter reported the first optical measurements [32]. By measuring the Shuster-Kubelka-Munk re-emission function and fluorescence of a vacuum dehydrated sample of silver clusters they presented and characterized for the very first time the excited electronic state of monodispersed atoms as well as optical absorption of collaterally produced clusters and particles. They assigned the broad absorption peaking at 3.3 eV to "intrazeolitic silver particles". The results triggered a large interest among scientist. Three years later Gellens *et al* reported the formation of the so called "color centers" in Ag-A zeolites, as a result of the reduction of Ag^+ ions upon dehydration. The authors assigned the process to the formation of linear Ag_3^{2+} clusters with Ag^0 located inside the sodalite cage opposite the framework D4R and in between two Ag^+ cations located at the D6R [33]. It was then believed that the yellow color is due to isolated cluster whereas the observed red color was associated to the formation of two or more interacting Ag_3^{2+} clusters in the sodalite cage to a maximum of four.

As stated above, the silicon-to-aluminum ratio is an important parameter of zeolite as this determines the number of exchangeable cations. In 1984, Johnson *and al.* detected distinct species following the reaction of sodium-exchanged zeolite Y with sodium, potassium, or rubidium vapor in dehydrated samples of zeolite Y [34]. Upon exposure to high concentrations of metal vapor he obtained stable ionic clusters (Na_4^{3+}, K_4^{3+} and Rb_4^{3+}) characteristic of small metallic particles located inside the zeolite cavities. Also interesting is that the researchers observed cluster species Ag_6^{q+} ($q = 1, 3, 5$) in which an unpaired electron is trapped at a cluster of six equivalent silver cations. In the same year ESR measurements on silver particles in zeolite A proved that during the reduction of silver-loaded zeolite A certain stages of cluster formation can be followed and detected the presence of Ag_6^{x+} and Ag_8^{y+} clusters which fit in the sodalite cage [35]. Two years later, Wang and Herron succeeded in encapsulation of CdS and PbS in zeolite Y matrix and indicated that zeolites can be thought as providing a solid solvent for this type of clusters [36]. Impressive work since, through simple experiments, they showed that their optical properties were dependent on size and state of aggregation, a property well-known already for metal particles. A fascinating conclusion was drawn: due to the fact that the transition from clusters to aggregates is not continuous but abrupt they suggested that aggregation inside the zeolite is a percolative process.

The first important review on the structure and chemistry of silver clusters in zeolites has been published by Sun and Self and the reader is encouraged to read this excellent work which inspired many scientists [9]. The spectroscopy and light induced processes of silver clusters in zeolites have been incipiently discussed by Chen *et al.* [37] For the first time they looked at photostimulated luminescence of silver-exchanged zeolite-Y. Once irradiated with 254 nm wavelength, the photoluminescence intensity of silver atoms centered at 505 nm decreases and a new absorption band shows up around 840 nm. By photostimulation of this absorption band, the fluorescence of silver atoms is observed and the photoluminescence intensity of silver atoms increases slightly. These phenomena were considered to be caused by the charge transfer between the zeolite framework and the entrapped silver atoms. The photostimulated luminescence of silver clusters encapsulated in zeolite-Y was caused by the recombination of luminescence centers with electrons released from their traps by photo- stimulation. Charge transfer from the framework oxygens to silver cations was also reported by Seff and Kim based on a color change from brick-red to yellow [38].

Following the pioneering work presented above, small metal particles started to attract the real interest of scientists due to their peculiar optical, electronic and catalytic activity that change from bulk properties to molecular-like properties once the size decreases under a certain range. For instance, a change in the electronic properties from a band structure to molecular orbitals levels was common for isolated species of a few atoms particle. Such change was almost invariable of the nature of the constituent atoms. Of a particular interest was to control the size and coordination because these would allow fine-tuning of electronic and luminescent properties of the clusters. However, to achieve that, a deep understanding of the electronic state was much desired. Based on such understanding the revealed structure-to-function relationship could finally open perspectives for practical applications.

The exact structure of metal clusters stabilized be the zeolite matrix remained under debate for a long time. Gellens et al. investigated the electronic properties of silver clusters using extended Hűckel framework [39]. They calculated isolated Ag_3 molecules in linear or nearly linear geometries and a strong similarity with the electronic spectrum of yellow Ag clusters were found. The proposed model for Ag3 encaged in the zeolite framework appeared to be weakly interacting with the zeolite lattice. Very interesting is that the scientists suggest the occurrence of an electron transfer of approximately two electrons from the cluster toward the zeolite framework increasing the charge density on the Si/Al model.

Until 1990s, simple structures as dimers and trimers of alkali-metal elements or noble metals were only experimentally investigated with UV–VIS and IR spectroscopy, Raman, ESR and EXAFS techniques [40–42]. These experiments indicated various shapes such as linear, bent or triangular structure and even interconversions between such structures were proposed. Precise studies of the effect of cluster shape, structure and interactions with the zeolite environment have been rather limited mainly due to a combination of instrumentation and techniques used. Other difficulties were found due to a distribution of particle size or problems with high mobility of precursors. Zeolite X and Y have been utilized as scaffolds to restrict and stabilize metal agglomeration. In such matrices, for instance, geometric models based on EXAFS experimental results revealed the formation of Pd_2, Pd_3 and Pd_4 clusters occupying adequate positions of the sodalite cages [43]. This showed again a "molecular" cluster elegantly stabilized in an open-framework zeolite. Texter *et al* encountered similar difficulties when investigated the formation of charged silver clusters of activated dehydration of zeolite A [44]. The nuclearity of these clusters formed in the sodalite unit was uncertain, believed to be between 6 and 14 while the dominant cluster had an absorption band centered at 2.72 eV and a higher band at 3.8 eV. Closely, more details about the cluster nuclearity and interactions with the framework have been brought via far-infrared experiments [45]. The cage vibrational mode characteristic for Ag_0 atoms isolated in zeolite Y has been identified at 89 cm^{-1} while silver-silver stretching modes were identified for zeolites A and Y encapsulated Ag_n^{q+} clusters with n = 2, 3 and 6. Via IR spectroscopy the presence of silver microcrystals located on the external surface of zeolite has been also demonstrated. Although important steps have been made, the structure of the metal cluster remains largely uncertain and very few have succeeded in providing strong evidences. Spectroscopic experiments, although widely qualitative, remain the basis in identifying a cluster species. Generally, larger clusters give rise to lower optical bandgap and inconsistencies in assignment of these bands were sometimes observed in literature.

A first attempt to establish the structure of the model Ag clusters in zeolite sample was made in 2004 by employing a combination of steady-state spectroscopy and K-edge XANES/EXAFS analyses [46]. The absorption spectra showed bands at 255 and 305 nm and from Ag K-edge EXAFS results, the structure of the cluster was presumably identified as Ag_4^{2+}. The coordination number 3.3 and the Ag-Ag distance of 2.7 Å suggest that the cluster consisted of 3 or 4 Ag atoms. The amount of the clusters increases with the Ag/Al ratio of Ag zeolites. Further spectroscopic investigations on very small particles Ag_2S and PbS in zeolite A showed photoluminescence in the visible range and lifetimes as long as 300 μs and these properties were shown to be strongly dependent on co-cations [47]. A broad investigation has been carried out for Li^+, Na^+, K^+, Rb^+, Cs^+, Mg^{2+}, Ca^{2+}, and Sr^{2+}. Interestingly, the unusual long luminescence lifetime has been interpreted as excitation energy transfer between Ag_2S and Ag_4S_2, a concept that comes often forward in literature.

3.2 State of the art

The long pathway laid and impressive work of scientists toward structural characterization of metal clusters provided a momentum wisely used by the recent scientists who combined time-resolved spectroscopy, X-ray excited optical luminescence (XEOL), electron spin resonance (ESR), X-ray crystallographic, Ag K-edge Extended X-ray Absorption Fine Structure (EXAFS) and density functional theory (DFT), time-resolved density functional theory (TD-DFT), X-ray photoelectron spectrometry (XPS), and used all the tools at hand to reveal with astonishing details the shape of these "molecules" and tag their luminescence properties [2–5, 11, 48–50].

A general and versatile synthesis protocol used for encapsulation of metal clusters within zeolite micropores enabled selective incorporation of a broad range of metals (Pt, Pd, Ir, Rh, and Ag) within NaA, a zeolite with channel windows too small for post synthesis encapsulation of metal precursors [51]. Encapsulation provides protection for clusters against sintering and contact with toxic impurities within environments. Better conditions of stabilization of Ag and Cu clusters in zeolite voids with formation of a self-assembled clusters system were found by Gurin *et al.* [52] As expected, zeolite voids could incorporate and stabilize clusters of smaller size than a void size. A few geometric structures have been proposed fitting erionite and mordenite voids (0.63–0.70 nm). The structures belong to the following point groups: Oh (cube), Td (tetrahedron), D4h (rectangular parallelepiped), and C2v (3D polyhedron). A stable geometry (rhombus) is found for Ag_4^+ in accordance with calculations presented earlier.

Restricted by the chapter size, we will further focus on the case of Ag clusters as these appear to be representative for these class of luminescent systems [49]. In this field, notable studies investigating the geometry and energetic properties of Ag_n clusters $3 \leq n \leq 6$ have been carried out using density functional theory

(DFT) calculations. FAU topologies predominantly accommodate Ag_2, Ag_3, Ag_4 and Ag_8 clusters while LTA frameworks prefer Ag_3 and Ag_6 clusters [3–5, 9, 33, 39, 53]. A number of geometries have been optimized inside a ZSM-5 zeolite whose ten-membered ring contains different numbers of Al atoms substituted for Si atoms of the SiO_2 framework. For Ag_5^+, a ditrigonal orthogonal geometry appeared as the most energetically stable configuration while a triangle Ag_3 cluster while for Ag_4 both a square and a tetragonal clusters have been considered. Ag_6 clusters inside ZSM-5(Al1) and ZSM-5(Al2) cavities preserve two stable configurations: a tetragon pair with a shared bond and edge and a pair with shared bonds [54]. DFT calculations also show that the cluster has two 5 s electrons populating the totally symmetric frontier orbitals, which leads to a stabilization of the cluster structure. The totally symmetric 5 s-based orbital corresponds to a super atom S-orbital. The optical transition modeled through time-dependent DFT calculations attributed the absorption peaks to an electronic transitions based on 5 s-type orbitals from the totally symmetric occupied orbital (S-orbital) to an unoccupied orbital with one node (P-orbital) [54].

The curious case of Ag_3 cluster. Experimental crystallographic data showed that Ag_3^+ and Ag_3^{2+} clusters form a linear configuration (with the later slightly bent) along the 3-fold axes through double six-rings of dehydrated zeolite X [55]. A weakly attractive interaction between Ag^+-Ag^+ could be concluded. Later, the assumption of a linear clusters has been confirmed by Zhao *et al* in a comprehensive series of DFT calculations [56]. The optimized geometries and binding energies of the most stable Ag^n, Ag^n -, Ag^n +, Ag^nH, Ag^nH- and Ag^nH+ with $2 \leq n \leq 7$ showed remarkable odd-even alternation behaviors. Silver behaves like an alkali-metal atom in the interaction between H and silver clusters. Surprisingly, in a different and fascinating DFT modeling study, the geometry of Ag_3 inside the void of ZSM-5 was demonstrated as a triangle and that the Ag-Ag orbital interaction as well as Ag-O electrostatic interactions determine such a different structure [57]. The structure is the same in both lower and high spin states, however, the high spin state leads to two types of triangles significantly distorted from the D3h configuration. The authors also established that the structural and electronic features are governed by the number of Al atoms trough Ag-Ag and Ag-O interactions. The modeling calculations support both a linear and a triangular structure [58].

Ag_4, a clear tetrahedron. Using already classical X-ray absorption measurements, the geometry of Ag_4 cluster has been carefully investigated in the FAU or LTA topologies [3–5, 48, 49, 59]. About 67% of Ag atoms constitute the oligomeric Ag_4 clusters and found located inside the sodalite cage [4]. Each Ag atom is bonded in to 2.2 water molecules and surrounded mostly by ca. 33% of Ag isolated cations located at the center of the S6R rings of the sodalite cage (18%) and in the center of the hexagonal prisms connecting the sodalite cages (15%). The fully Ag-exchanged sample FAUY contains a similar Ag_4 geometry, with the difference that the cation distribution is slightly changed with of Ag_4 cluster surrounded by ca. 25% of silver ions located close to the center of the S6Rs and ca. 13% of silver atoms located in the center of the hexagonal prisms. This excellent study is particularly interesting for its extensive characterization which indicates a strong and direct influence of silver loading and host environment on the cluster ionization potential. This new finding is also correlated to the cluster's optical and structural properties. By fine-tuning of the zeolite environment the researches achieved clusters with photoluminescence quantum yield approaching unity. Another Ag_4 cluster (dehydrated Ag_1Li_{11}-LTA zeolite) was found to feature a remarkable EQE of 83% with an emission maxima around 545 nm when excited between 300 and 400 nm [48]. The presence of Li^+ clearly changes the luminescence properties of this cluster in two ways: by shielding the interaction between cluster and zeolite oxygen framework and by contraction of the lattice parameter leading to a tightly confined cluster inside the sodalite cage of LTA zeolite.

The octahedron cluster, Ag_6. Using a different, Ag clusters in LTA and FAU zeolites have been characterized via luminescence properties which were shown before to depend on the nature of the co-cation, the amount of exchanged silver, and the host topology. A broad pallet of emissive species Ag_n,Na-X and Ag_n,Na-Y, $1 \leq n \leq 12$ were observed with spectral properties ranging from 380 to 700 nm and further used as "fingerprint" in cluster type identification [3]. The luminescence decays on a time scale ranging from ns to μs and the transitions were attributed to excited state processes involving spin forbidden transitions and intersystem crossing. A singlet-triplet transition in the case of $Ag_6{}^{2+}$ cluster or a doublet-quadruplet transition in the case of $Ag_6{}^{+}$ have been suggested. Similar long luminescence decay times were also found in other types of silver-zeolite systems. The exact structure of each luminescent species and the nature of the luminescent electronic state still remain a subject of investigations. The structure of hexasilver molecule $Ag_6{}^0$ has been further resolved by the group of Seff *et al* via single crystal diffraction experiments and showed that these crystals contains octahedral subunits [60]. $Ag_6{}^0$ had formed in varying amounts in up to 60% in the sodalite cavities of the crystals studied. The hexasilver was shown to be surrounded by eight Ag^+ ions distributed in a cub, one in each face of the octahedron. In addition, the presence of even larger clusters is $Ag_8{}^+$ or $Ag_{14}{}^{8+}$ was detected.

4. Electronic properties and the origin of intense luminescence

Validated by various research groups, the photophysical properties of metal clusters encaged in zeolite matrices are strongly dependent on numerous factors like ligand, spatial confinement, charge state, water content, electrostatic charge of the cavity or co-cation type [11]. Perhaps one of the largest scientific debates among scientist in this field is understanding the size dependent properties of a metal cluster [9]. Although the strong quantum confinement of the electrons separating the continues density of states into discrete energy levels is generally recognized as the origin of the optical transition, the fundamental photophysical mechanisms underlying their emission are poorly understood [61]. Metal clusters in various size regimes display molecule-like optical characteristics featuring HOMO-LUMO bandgap with transition to metal properties at high nuclearity. Obtaining well-defined clusters often remains questionable in designing nanostructures with specific functional properties. An understanding of both structural and electronic features by invoking their familiar aspects like the discrete electronic shell has led to a few concepts which attempt to rule the characteristics of the molecular-like entity. Substantial theoretical and computational efforts were made to understand and predict the fundamental properties associated with the existing and emerging metal clusters and develop a general valid theory [4, 7, 16].

In the examples presented earlier, the intense or dim luminescence has been attributed to various phenomena occurring in the excited state, effects mainly related to intersystem crossing (leading to phosphorescence), charge transfer and recombination, ion migration or structural changes [12, 62, 63]. Intriguingly enough, limited time-resolved spectroscopy experiments have shown that the luminescence decays on time scales ranging from fs to ms, a characteristic similar to molecules undergoing a complex excited-state dynamics. Several kinetic schemes have been proposed indicating luminescence either as a spin forbidden radiative transition to the ground state (phosphorescence) or recombination of electrons and holes trapped in the zeolite matrix (fluorescence) [2, 13, 59, 64].

The electronic configuration of the main constituent atoms Cu, Ag, and Au is situated at 3d, 4d, and 5d period, respectively. When a silver atoms with an electronic configuration of $4d^{10}5s^1$ forms a cluster, the electronic properties are determined by the frontier orbitals resulted from a linear combination of the 5 s orbitals [50, 54]. The 5 s orbitals can be regarded within the concept of "super atom" in analogy to the orbitals in hydrogen-like atoms. In gas-phase, clusters are assumed to behave like giant atoms and obey the same rules as atoms. Similarly, principles like orbital hybridization or Hund's rule can be applied. Inspired by the Jellium model, the confined electrons move within a smeared positive region that is evenly distributed over about the cluster volume [65]. In this model, the frontier orbitals with different numbers of nodes can be classified as S, P, and D and have an angular momentum quantum number L = 0, 1, and 2 for the S, P, and D orbitals, respectively) corresponding to the number of nodes of the 5 s-based orbitals. [67,72] In this light, the magic number of bare silver clusters and the selection rule in their electronic transitions ($\Delta L = \pm 1$) can be resolved [7, 66].

Utilizing the super-atom orbital concept to understand properties of silver clusters inside ZSM-5 zeolite, Yumura *et al.* investigated the energetic properties of Ag_n clusters by using DFT and TD-TDF calculations [54]. TD- DFT optimized geometries of Ag_n–ZSM-5(Al_m), where $3 < n < 6$ and $1 < m < 5$, showed an intense oscillator strength at the electronic transition between 5 s-based orbitals from the totally symmetric orbital (S-orbital) to that with one node (P-orbital). The S → P electronic transitions obeys the selection rule for cluster in gas-phase. Previously, DFT calculations of the Ag_3 and Ag_4 clusters inside a 10-membered ring of the ZSM-5 zeolite showed that photon absorption is due to a transitions from a completely symmetric 5 s-based orbital to a 5 s-based orbital with one node [67]. The absorptions spectra are "modulated" or strongly affected by the encapsulations as this induces cluster distortion leading to interactions between clusters and the framework oxygen atoms. The optical properties of Ag_4 and Ag_6 encapsulated inside the sodalite cavity of LTA zeolite were investigated using similar DFT and TD-DFT methods by Cuong *et al.* [50] Hydrated quadruply charged silver hexamer features a strong absorption band at 420 nm which is very sensitive to its charge. In the case of hydrated doubly charged silver tetramer cluster, the absorption band shifts slightly and steadily to lower energy with the increasing amount of interacting water molecules. The presence of water molecules pushes the silver tetramer away from the cavity center. Water molecules take the role as ligands and induce a splitting of the energy levels of excited states of both $Ag_4{}^{2+}$ and $Ag_6{}^{4+}$ clusters. As we will see bellow, this splitting causes the optical properties of the clusters to change significantly.

In a remarkable study, Grandjean *et al* investigated the structure and electronic properties at the origin of the luminescence of Ag_4 clusters confined in Ag-LTA-zeolite by a unique combination of XEOL-EXAFS, (TD)-DFT calculations and time-resolved spectroscopy [5]. XEOL experiments showed that the species at the origin of the bright green luminescence observed in Ag_3K_9-exchanged LTA zeolites are Ag_4 clusters with short Ag-Ag distances of 2.82 Å in which each Ag atom is bonded in average to 2 water molecules at 2.36 Å. This results suggested the presence of two isomers $Ag_4(H_2O)_4$ and $Ag_4(H_2O)_2$ clusters with a 40/60 ratio in this Ag3K9-LTA sample. DFT calculations confirmed the presence of the two stable isomers. The best agreement between both the calculated structures and absorption spectra with those measured experimentally were obtained when applying a + 2 charge preferentially localized on the Ag_4 cluster. The calculated frontier orbitals for both $Ag_4(H_2O)_4{}^{2+}$ and $Ag_4(H_2O)_2{}^{2+}$ isomers are a superposition of 50% from Ag 5 s atomic orbitals and of up to 25% from the oxygen states of the surrounding framework oxygens and water molecules. The lowest cluster configuration is formed from a doubly occupied $^1S_0$ HOMO of totally symmetric s-type and two sets of three singlet 1P and three triplet 3P LUMOs of p-type character with one node which corresponds to the assumed two electron model cluster. The coordination with water molecules lifts the degeneracy of the LUMOs orbitals. As a result LUMOs in $Ag_4(H_2O)_4^{2+}$ and $Ag_4(H_2O)_2^{2+}$ clusters are split into six excited states i.e. three singlet $^1P(S = 0, L = 1, m_l = -1, +1 \text{ or } 0)$ and three triplet $^3P(S = 1, L = 1, m_l = -1, +1 \text{ or } 0)$ states. Due to the quasi isoenergetic position of the high-energy triplet $^3P(S = 1, L = 1, m_l = 0)$ state with the ^1P singlet states corroborated to a silver large *spin-orbit* coupling, an intersystem crossing takes place. Photon absorption promotes the clusters in the excited singlet state. A fraction of ^1P singlet states population is transferred to the high-lying triplet state that finally decays via internal conversion into the low-lying $^3P(S = 1, L = 1, m_l = -1 \text{ or } +1)$ triplet state. Although a spin forbidden process, the bright green emission takes place as a radiative transition from the lowest triplet excited state $^3P(S = 1, L = 1, m_l = -1)$ to the ground singlet state $^1S_0$.

5. Conclusions and outlook

Even though the super atom model could provide a qualitative description of the luminescence properties, the exact picture to contain the dynamics and kinetics of electron transition of the excited state remains blurred. By judiciously manipulating the interactions at the interface between cluster and zeolite framework together with a careful calculation of time-dependent density functional theory, a few proposed kinetic scheme claim to elucidate the long-lived emitting state. Simply considering an excited state intersystem crossing process followed by a radiative decay to the ground state would not fully explain the presence of a complex array of decay components observed on a fs and ps time scales, decay observed in the photophysics of many clusters. Ultrafast decay components would indeed indicate the occurrence of an intersystem crossing phenomena but can also suggest internal conversion or fast structural relaxation which are so common for molecules. The ns components would suggest decays between states of the same spin multiplicity, like fluorescence. Are these components considered at all when concluding the excited state dynamics? Why do these emissions indicate a Stokes shift as large as 12–15000 cm^{-1}? Are there other intermediate states that are not yet experimentally revealed? When all is said and done, are all the fruits from the tree harvested?

As shown above, the lack of full understanding of the photophysics of metal clusters resides in its optical properties which depend on so many factors including cluster size, temperature, surface ligands geometry, cluster assembly structure, humidity, etc. One think is clear. Dedicated time-resolved spectroscopy techniques like transient absorption and fluorescence up-conversion at the femtosecond scale should come more consistently into play to provide precise and reliable information. Only then a complete picture of the relaxation and decay pathways of excited electrons immediately after the population of the Franck-Condon state can be achieved. More modern time-resolved spectroscopy techniques, like for instance X-ray diffraction based on X-Ray Free-Electron Laser (XFEL) methods, could elucidate changes in structural dynamics immediately after excitation. Most importantly, with respect to the paramount importance, universality and complexity of the model, the experimental part, be that spectroscopy, diffraction or ESR techniques, should take the first place in providing the arguments while modeling and computation could shortly support the experimental findings. Only when a complete picture of the excited state processes is achieved one can tune the electronic structure and thus interfere with the optical properties of the metals clusters. Hence, the advance achieved will be of immediate interest to a broader pool of researchers and will open real pathways for practical applications.

Author details

Eduard Fron
Director Core Facility for Advanced Spectroscopy, KU Leuven, Molecular Imaging and Photonics, Leuven, Belgium

*Address all correspondence to: eduard.fron@kuleuven.be

References

[1] Awala, H. et al. Template-free nanosized faujasite-type zeolites. *Nature Materials* **14**, 447-451 (2015).

[2] Calzaferri, G., Leiggener, C., Glaus, S., Schurch, D. & Kuge, K. The electronic structure of Cu+, Ag+, and Au+ zeolites. *Chemical Society Reviews* **32**, 29-37 (2003).

[3] De Cremer, G. et al. Characterization of Fluorescence in Heat-Treated Silver-Exchanged Zeolites. *Journal of the American Chemical Society* **131**, 3049-3056 (2009).

[4] Fenwick, O. et al. Tuning the energetics and tailoring the optical properties of silver clusters confined in zeolites. *Nature Materials* **15**, 1017-1022 (2016).

[5] Grandjean, D. et al. Origin of the bright photoluminescence of few-atom silver clusters confined in LTA zeolites. *Science* **361**, 686-689 (2018).

[6] Huang, Z. et al. 3D-3D topotactic transformation in aluminophosphate molecular sieves and its implication in new zeolite structure generation. *Nat Commun* **11**, 3762 (2020).

[7] Knight, W.D. et al. ELECTRONIC SHELL STRUCTURE AND ABUNDANCES OF SODIUM CLUSTERS. *Physical Review Letters* **52**, 2141-2143 (1984).

[8] Mintova, S., Jaber, M. & Valtchev, V. Nanosized microporous crystals: emerging applications. *Chemical Society Reviews* **44**, 7207-7233 (2015).

[9] Sun, T. & Seff, K. SILVER CLUSTERS AND CHEMISTRY IN ZEOLITES. *Chemical Reviews* **94**, 857-870 (1994).

[10] Tao, Y.S., Kanoh, H., Abrams, L. & Kaneko, K. Mesopore-modified zeolites: Preparation, characterization, and applications. *Chemical Reviews* **106**, 896-910 (2006).

[11] Hashimoto, S. Zeolite photochemistry: impact of zeolites on photochemistry and feedback from photochemistry to zeolite science. *Journal of Photochemistry and Photobiology C-Photochemistry Reviews* **4**, 19-49 (2003).

[12] Zhang, G.H., Liu, X.S. & Thomas, J.K. Radiation induced physical and chemical processes in zeolite materials. *Radiation Physics and Chemistry* **51**, 135-152 (1998).

[13] Chen, W., Joly, A.G. & Roark, J. Photostimulated luminescence and dynamics of AgI and Ag nanoclusters in zeolites. *Physical Review B* **65** (2002).

[14] Heo, N.H., Kim, H.S., Lim, W.T. & Seff, K. Synthesis and crystal structure of Ag4I4 nanoclusters in the sodalite cavities of fully K+-exchanged zeolite A. *Journal of Physical Chemistry B* **108**, 3168-3173 (2004).

[15] Lim, W.T. et al. Synthesis and crystal structure of Ag4Br4 nanoclusters in the sodalite cavities of fully K+-exchanged zeolite A (LTA). *Bulletin of the Korean Chemical Society* **26**, 1090-1096 (2005).

[16] Joshi, C.P., Bootharaju, M.S. & Bakr, O.M. Tuning Properties in Silver Clusters. *Journal of Physical Chemistry Letters* **6**, 3023-3035 (2015).

[17] Wang, S.B. & Peng, Y.L. Natural zeolites as effective adsorbents in water and wastewater treatment. *Chemical Engineering Journal* **156**, 11-24 (2010).

[18] Perez-Ramirez, J., Christensen, C.H., Egeblad, K. & Groen, J.C. Hierarchical zeolites: enhanced utilisation of microporous crystals in catalysis by advances in materials

design. *Chemical Society Reviews* **37**, 2530-2542 (2008).

[19] Olsbye, U. et al. Conversion of Methanol to Hydrocarbons: How Zeolite Cavity and Pore Size Controls Product Selectivity. *Angewandte Chemie-International Edition* **51**, 5810-5831 (2012).

[20] Davis, M.E. & Lobo, R.F. ZEOLITE AND MOLECULAR-SIEVE SYNTHESIS. *Chemistry of Materials* **4**, 756-768 (1992).

[21] Cundy, C.S. & Cox, P.A. The hydrothermal synthesis of zeolites: Precursors, intermediates and reaction mechanism. *Microporous and Mesoporous Materials* **82**, 1-78 (2005).

[22] Cundy, C.S. & Cox, P.A. The hydrothermal synthesis of zeolites: History and development from the earliest days to the present time. *Chemical Reviews* **103**, 663-701 (2003).

[23] Kasai, P.H. ELECTRON SPIN RESONANCE STUDIES OF GAMMA- AND X-RAY-IRRADIATED ZEOLITES. *Journal of Chemical Physics* **43**, 3322-& (1965).

[24] *Cejka, J.r. Introduction to zeolite science and practice, Edn. 3rd rev. (Elsevier, Amsterdam ; Boston; 2007).

[25] Breck, D.W. Zeolite molecular sieves: structure, chemistry, and use. (Wiley, New York,; 1973).

[26] Breck, D.W. Zeolite molecular sieves : structure, chemistry, and use. (R.E. Krieger, Malabar, Fla.; 1984).

[27] Dyer, A. An introduction to zeolite molecular sieves. (J. Wiley, Chichester ; New York; 1988).

[28] Preuss, E., Linden, G. & Peuckert, M. MODEL-CALCULATIONS OF ELECTROSTATIC FIELDS AND POTENTIALS IN FAUJASITE TYPE ZEOLITES. *Journal of Physical Chemistry* **89**, 2955-2961 (1985).

[29] Baretz, B.H. & Turro, N.J. FLUORESCENCE STUDIES OF PYRENEALDEHYDE ADSORBED WITHIN ZEOLITE SUPERCAGES. *Journal of Photochemistry* **24**, 201-205 (1984).

[30] Liu, X., Iu, K.K. & Thomas, J.K. PHOTOPHYSICAL PROPERTIES OF PYRENE IN ZEOLITES. *Journal of Physical Chemistry* **93**, 4120-4128 (1989).

[31] Liu, X. & Thomas, J.K. FORMATION AND PHOTOPHYSICAL PROPERTIES OF CDS IN ZEOLITES WITH CAGES AND CHANNELS. *Langmuir* **5**, 58-66 (1989).

[32] Kellerman, R. & Texter, J. OPTICAL-ABSORPTION OF SILVER ATOMS AND SILVER CLUSTERS IN ZEOLITE-Y. *Journal of Chemical Physics* **70**, 1562-1563 (1979).

[33] Gellens, L.R., Mortier, W.J. & Uytterhoeven, J.B. ON THE NATURE OF THE CHARGED SILVER CLUSTERS IN ZEOLITES OF TYPE-A, TYPE-X AND TYPE-Y. *Zeolites* **1**, 11-18 (1981).

[34] Harrison, M.R. et al. IONIC AND METALLIC CLUSTERS OF THE ALKALI-METALS IN ZEOLITE-Y. *Journal of Solid State Chemistry* **54**, 330-341 (1984).

[35] Grobet, P.J. & Schoonheydt, R.A. ELECTRON-SPIN-RESONANCE ON SILVER CLUSTERS IN ZEOLITE-A. *Surface Science* **156**, 893-898 (1985).

[36] Wang, Y. & Herron, N. OPTICAL-PROPERTIES OF CDS AND PBS CLUSTERS ENCAPSULATED IN ZEOLITES. *Journal of Physical Chemistry* **91**, 257-260 (1987).

[37] Chen, W., Wang, Z.G., Lin, Z.J., Xu, Y. & Lin, L.Y. Photoluminescence

of ZnS clusters in zeolite-Y. *Journal of Materials Science & Technology* **13**, 397-404 (1997).

[38] Kim, Y. & Seff, K. STRUCTURE OF A VERY SMALL PIECE OF SILVER METAL - OCTAHEDRAL AG6 MOLECULE, 2 CRYSTAL-STRUCTURES OF PARTIALLY DECOMPOSED VACUUM-DEHYDRATED FULLY AG+-EXCHANGED ZEOLITE-A. *Journal of the American Chemical Society* **99**, 7055-7057 (1977).

[39] Gellens, L.R., Mortier, W.J., Lissillour, R. & Lebeuze, A. ELECTRONIC-STRUCTURE OF THE SILVER CLUSTERS IN ZEOLITES OF TYPE-A AND THE FAUJASITE TYPE BY MOLECULAR-ORBITAL CALCULATIONS. *Journal of Physical Chemistry* **86**, 2509-2516 (1982).

[40] Jansson, K. & Scullman, R. OPTICAL-ABSORPTION SPECTRA OF PTO AND PT-2 IN RARE-GAS MATRICES. *Journal of Molecular Spectroscopy* **61**, 299-312 (1976).

[41] Hulse, J.E. & Moskovits, M. INTERACTION OF CO WITH MATRIX-ISOLATED NI CLUSTERS - MODEL FOR CO CHEMISORBED ON NI. *Surface Science* **57**, 125-142 (1976).

[42] Montano, P.A., Schulze, W., Tesche, B., Shenoy, G.K. & Morrison, T.I. EXTENDED X-RAY-ABSORPTION FINE-STRUCTURE STUDY OF AG PARTICLES ISOLATED IN SOLID ARGON. *Physical Review B* **30**, 672-677 (1984).

[43] Moller, K., Koningsberger, D.C. & Bein, T. STABILIZATION OF METAL ENSEMBLES AT ROOM-TEMPERATURE - PALLADIUM CLUSTERS IN ZEOLITES. *Journal of Physical Chemistry* **93**, 6116-6120 (1989).

[44] Texter, J., Kellerman, R. & Gonsiorowski, T. FORMATION OF CHARGED SILVER CLUSTERS AND THEIR REVERSIBLE SILVER ION DESORPTION IN ZEOLITE A. *Journal of Physical Chemistry* **90**, 2118-2124 (1986).

[45] Baker, M.D., Ozin, G.A. & Godber, J. FAR-INFRARED STUDIES OF SILVER ATOMS, SILVER IONS, AND SILVER CLUSTERS IN ZEOLITE-A AND ZEOLITE-Y. *Journal of Physical Chemistry* **89**, 305-311 (1985).

[46] Shibata, J. et al. Structure of active Ag clusters in Ag zeolites for SCR of NO by propane in the presence of hydrogen. *Journal of Catalysis* **227**, 367-374 (2004).

[47] Leiggener, C. & Calzaferri, G. Synthesis and luminescence properties of Ag2S and PbS clusters in zeolite A. *Chemistry-a European Journal* **11**, 7191-7198 (2005).

[48] Baekelant, W. et al. Luminescent silver-lithium-zeolite phosphors for near-ultraviolet LED applications. *Journal of Materials Chemistry C* 7, 14366-14374 (2019).

[49] Coutino-Gonzalez, E., Roeffaers, M. & Hofkens, J. in Dyes and Photoactive Molecules in Microporous Systems, Vol. 183. (eds. V. MartinezMartinez & F.L. Arbeloa) 75-1032020).

[50] Cuong, N.T., Nguyen, H.M.T., Phuong, P.H.M. & Nguyen, M.T. Optical properties of the hydrated charged silver tetramer and silver hexamer encapsulated inside the sodalite cavity of an LTA-type zeolite. *Physical Chemistry Chemical Physics* **18**, 18128-18136 (2016).

[51] Choi, M., Wu, Z.J. & Iglesia, E. Mercaptosilane-Assisted Synthesis of Metal Clusters within Zeolites and Catalytic Consequences of Encapsulation. *Journal of the American Chemical Society* **132**, 9129-9137 (2010).

[52] Gurin, V.S., Bogdanchikova, N.E. & Petranovskii, V.P. Self-assembling of silver and copper small clusters within the zeolite cavities: prediction of geometry. *Materials Science & Engineering C-Biomimetic and Supramolecular Systems* **18**, 37-44 (2001).

[53] Fonseca, A.M. & Neves, I.C. Study of silver species stabilized in different microporous zeolites. *Microporous and Mesoporous Materials* **181**, 83-87 (2013).

[54] Yumura, T., Kumondai, M., Kuroda, Y., Wakasugi, T. & Kobayashi, H. Utilizing super-atom orbital ideas to understand properties of silver clusters inside ZSM-5 zeolite. *Rsc Advances* **7**, 4950-4959 (2017).

[55] Lee, S.H., Kim, Y. & Seff, K. Weak Ag+-Ag+ bonding in zeolite X. Crystal structures of Ag92Si100Al92O384 hydrated and fully dehydrated in flowing oxygen. *Microporous and Mesoporous Materials* **41**, 49-59 (2000).

[56] Zhao, S., Liu, Z.P., Li, Z.H., Wang, W.N. & Fan, K.N. Density functional study of small neutral and charged silver cluster hydrides. *Journal of Physical Chemistry A* **110**, 11537-11542 (2006).

[57] Yumura, T., Nanba, T., Torigoe, H., Kuroda, Y. & Kobayashi, H. Behavior of Ag-3 Clusters Inside a Nanometer-Sized Space of ZSM-5 Zeolite. *Inorganic Chemistry* **50**, 6533-6542 (2011).

[58] Janata, E. Structure of the trimer silver cluster Ag-3(2+). *Journal of Physical Chemistry B* **107**, 7334-7336 (2003).

[59] Fron, E. et al. Structural and Photophysical Characterization of Ag Clusters in LTA Zeolites. *Journal of Physical Chemistry C* **123**, 10630-10638 (2019).

[60] Heo, N.H., Kim, Y., Kim, J.J. & Seff, K. Surprising Intrazeolitic Chemistry of Silver. *Journal of Physical Chemistry C* **120**, 5277-5287 (2016).

[61] Chan, B. Re-examining the electronic structure of fluorescent tetra-silver clusters in zeolites. *Physical Chemistry Chemical Physics* **23**, 1984-1993 (2021).

[62] Ngo Tuan, C., Hue Minh Thi, N. & Minh Tho, N. Theoretical modeling of optical properties of Ag-8 and Ag-14 silver clusters embedded in an LTA sodalite zeolite cavity. *Physical Chemistry Chemical Physics* **15**, 15404-15415 (2013).

[63] Keirstead, A.E., Schepp, N.P. & Cozens, F.L. Influence of the alkali metal cation on the distance of electron migration in zeolite Y: A nanosecond laser photolysis study. *Journal of Physical Chemistry C* **111**, 14247-14252 (2007).

[64] Przystawik, A., Radcliffe, P., Goede, S.G., Meiwes-Broer, K.H. & Tiggesbaeumker, J. Spectroscopy of silver dimers in triplet states. *Journal of Physics B-Atomic Molecular and Optical Physics* **39**, S1183-S1189 (2006).

[65] Martins, J.L., Car, R. & Buttet, J. VARIATIONAL SPHERICAL MODEL OF SMALL METALLIC PARTICLES. *Surface Science* **106**, 265-271 (1981).

[66] Chou, M.Y., Cleland, A. & Cohen, M.L. TOTAL ENERGIES, ABUNDANCES, AND ELECTRONIC SHELL STRUCTURE OF LITHIUM, SODIUM, AND POTASSIUM CLUSTERS. *Solid State Communications* **52**, 645-648 (1984).

[67] Yumura, T. et al. Combined Experimental and Computational Approaches To Elucidate the Structures of Silver Clusters inside the ZSM-5 Cavity. *Journal of Physical Chemistry C* **118**, 23874-23887 (2014).

Chapter 8

The Effect of HNO_3 and/or NaOH Treatments on Characteristics of Mordenite

Wega Trisunaryanti, Karna Wijaya, Desi Suryani and Uswatul Chasanah

Abstract

Modification of mordenite had been performed by HNO_3 and/or NaOH treatments and used as a catalyst in the hydrotreating of pyrolyzed α-cellulose. The Si/Al ratio of mordenites was analyzed by Inductively Coupled Plasma (ICP), the acidity was gravimetrically determined using ammonium vapor adsorption, the crystallinity was identified by X-ray Diffraction (XRD), and pore characters of mordenites (BHM, $BAM_{0.1}$, and $BAM_{0.5}$) were analyzed using Gas Sorption Analyzed (GSA). Catalytic activity and selectivity of the mordenites were evaluated in the hydrotreating of pyrolyzed α-cellulose under 20 mL minute^{-1} H_2 gas flow at 450 °C for 2 h with the weight ratio of catalyst:feed was 1:60. The liquid products were analyzed using Gas Chromatography-Mass Spectrometer (GC–MS). The acidity of mordenites decreased along with the increase of Si/Al rasio, except for the $AM_{0.1}$ mordenite. The average pore diameter of BHM, $BAM_{0.1}$ and $BAM_{0.5}$ was 2.96; 3.34 and 4.53 nm, respectively. The $BAM_{0.1}$ showed the highest catalytic activity in producing liquid fraction (64.04 wt%). The $BAM_{0.5}$ showed the highest catalytic selectivity towards biofuels, i.e. 1-pentene 0.44 (wt%); 2-heptyne 2.75 (wt%) and 1-propanol 3.05 (wt%) from the hydrotreating process of pyrolyzed α-cellulose.

Keywords: modification, mordenite, HNO_3 treatment, NaOH treatment, hydrotreating

1. Introduction

Zeolite is a crystalline alumino-silicate material that has a micropore structure. The three-dimensional structure of the zeolite is composed of AlO_4 and SiO_4 which are related to each other by sharing electrons from oxygen and are arranged tetrahedral [1]. Zeolites have been widely used in industrial processes as environmentally friendly heterogeneous catalysts, ion exchange and adsorbents due to their high specific surface area, large pore volume, uniform micropore channels and excellent thermal and hydrothermal stability. The use of zeolite as a catalyst in various industries is limited due to its narrow pores. The reactant components that have a large molecular size will experience difficulties during the mass transfer process which will affect the catalytic activity [2, 3]. Increasing the diffusion of the reactants on the catalyst can be done by increasing the pore size of the zeolite, or by creating a mesoporous structure on the zeolite. Therefore, the meso-sized catalyst structure is preferred by most industries, especially the petroleum processing industry because it can increase the effectiveness in the petroleum processing process [4].

Researchers [5] have synthesized mesoporous zeolite-Y using a mold in the form of pluronic F_{127}. In general, the mold used is a surfactant that has a long chain. This method has been effective in making of mesoporous zeolite-Y [6].

Mesoporous which is more practical and economical, namely the alkaline treatment known as desilication. This method has been used by many researchers, one of them is researchers [7] that have succeeded in synthesizing mesoporous mordenite using NaOH. Mesoporous distribution in the mordenite increased significantly after being treated with NaOH. This shows that NaOH treatment can produce mesoporous zeolites with a fairly practical and inexpensive process.

The Si/Al ratio in zeolite is one of the characteristics that is closely related to this method of mesoporosity recognition. Alkaline treatment can cause dominant desilication if the Si/Al ratio is above 25 [7]. The low Si/Al ratio value has a limited effect from the alkaline treatment, this is due to the high aluminum content in the zeolite framework, so to get the optimal Si/Al ratio for dominant desilication, the dealumination process needs to be done first. Researchers [8] have succeeded in dealing with H-mordenite, zeolite-Y and ZSM using nitric acid. The dealumination can increase the Si/Al ratio without significantly impairing the crystallinity. In addition, previous study [9] used nitric acid to alluminate zeolite HMCM-22. Nitric acid is considered a strong acid to increase the Si/Al ratio of the zeolite. The dealumiation of mordenite with nitric acid showed an increase in the Si/Al ratio without destroying its crystallinity. This study was conducted by [10].

Mordenite is the most widely used catalyst in industrial processes, one of which is the petrochemical and petroleum process [11]. Mordenite has high thermal stability and is more resistant to acids. Mordenite is also a catalyst with a stronger acid site compared to zeolite-Y [12]. Therefore, mordenite can be used to produce biofuel. One of the raw materials in making biofuel is cellulose.

The research of [13] stated that cellulose is a candidate biomass source with the most potential to replace fossil fuels in the next few years, because it can reduce greenhouse gas emissions than fossil fuels. One type of cellulose polymer is α-cellulose. The α-cellulose polymer has a degree of polymerization above 200 [14]. This causes α-cellulose to be a carbon source that can be converted into biofuel. A-cellulose can be converted into fuel fractions in the form of hydrocarbons and alcohol by using a pyrolysis process.

However, the pyrolysis process is not effective in producing hydrocarbon and alcohol fractions, this is because bio-oil from pyrolysis is known to have acidic properties, is unstable for heating, and contains many oxygenated compounds [15]. Therefore, one effort that can be done to improve the quality of bio-oil from hydrolyzed α-cellulose is the hydrotreating process. The hydrotreating process can be applied to produce biofuels. This causes the hydrotreating process to become a major concern in the petroleum processing industry because it can produce products that have better selling power and usability [16].

Based on the description above, mordenite modification is carried out by treating HNO_3 and/or NaOH. This study examines the effect of HNO_3 and/or NaOH treatment on Si/Al ratio, crystallinity and acidity of mordenite and the effect of NaOH treatment on mordenite pore character. Then the activity and selectivity of the catalyst were tested in the hydrolyzed α-cellulose hydrotreating process.

2. Experimental methods

2.1 Materials and methods

2.1.1 Materials

The materials used in this study were synthetic mordenite type zeolite (HSZ-640HOA, Tosoh Corporation Japan®), NH_4Cl, NaOH, 37% HCl, 40% HF, 65% HNO_3, and 25% NH_3 (Merck®). Besides, cellulose (Sigma-Aldrich #C8002), Whatman filter paper No.42, bi-distilled water, N_2, and H_2 gas (PT. Samator Gas) were utilized.

2.1.2 Instrumentations

The instruments used for analysis are X-ray Diffraction (Rigaku® Miniflex 600), Inductively Coupled Plasma (Shimadzu® ICPE-9820), Surface Area Analyzer (Quantachrome® NOVAtouch), and Gas Chromatography–Mass Spectrometer (Shimadzu® QP 2010S).

2.2 Procedures

2.2.1 Acid treatment on mordenite

Ion exchange of zeolite type mordenite (HSZ-604HOA) was carried out with 1 M NH_4Cl for 3 hours at 70 °C and calcined in a furnace at 550 °C for 5 hours as HM. Zeolite HM formed was refluxed in a solution of HNO_3 (0.1 and 0.5) at a temperature of 70 °C for 2 hours, then washed with bi-distilled water until it was neutral, dried overnight at 110 °C, then calcined in a furnace 550 °C for 5 hours to form AM-0.1 and AM-0.5.

2.2.2 Alkaline treatment on mordenite

The HM; AM-0.1; and AM-0.5 were stirred respectively in 0.1 M NaOH for 15 minutes at room temperature (27°C), then ion exchange was carried out with 1 M NH_4Cl solution at 70 °C for 2 hours, washed with bi-distilled water until it was neutral, dried overnight, then calcined in a furnace at a temperature of 550 °C for 5 hours. In the end, the zeolite BHM; BAM-0.1, and BAM-0.5 were obtained.

2.2.3 Pyrolysis of α-cellulose

A total of 25 grams of α-cellulose is put into the pyrolysis reactor which is made of stainless steel. Pyrolysis was carried out with N_2 gas with a flow rate of 20 mL min^{-1} at a temperature of 600 °C for 3 hours. The resulting liquid product is cooled by a condenser and collected in a flask. The pyrolysis product of α-cellulose was then analyzed using a Gas Chromatography–Mass Spectrometer (GC–MS) to determine its compound content.

2.2.4 Catalyst activity test

Hydrotreating reaction of α-cellulose in three variations of conditions, namely thermal hydrotreating, catalytic hydrotreating with mordenite catalyst before and after NaOH alkaline treatment. The hydrotreating reaction with a mordenite catalyst was carried out on cellulose with a catalyst/feed ratio of 1/60 (w/w). The feed and catalyst are put into the sleeve with the feed position under the catalyst, then the sleeve is inserted into the reactor. The hydrotreating reaction was carried out with hydrogen gas with a flow rate of 20 mL min^{-1} at a temperature of 450 °C for 2 hours. In comparison, a thermal hydrotreating reaction is also carried out. The product formed is flowed through a condenser (cooler) to a flask which is cooled with an ice bath.

The percentage (%) conversion of the hydrotreating reaction results is determined by the following formula:

$$\text{Liquid product}(\text{wt.\%}) = \frac{M_{j1} - M_{j0}}{M_s - (M_{u1} - M_{u0})} \times 100\% \quad (1)$$

$$\text{Coke}(\text{wt.\%}) = \frac{M_{k1} - M_{k0}}{M_s - (M_{u1} - M_{u0})} \times 100\% \quad (2)$$

$$\text{Gas}(\text{wt.\%}) = 100\% - (\text{liquid product} + \text{coke}) \quad (3)$$

Where M_s = mass of α-cellulose; M_{j0} = mass of an empty flask; M_{j1} = flask mass after hydrotreating; M_{u0} = mass of the feed container before hydrotreating; M_{u1} = mass of the feed container after hydrotreating; M_{k0} = mass of catalyst before hydrotreating; M_{k1} = mass of catalyst after hydrotreating.

3. Results and discussion

3.1 The effect of HNO_3 and/or NaOH treatment on Si/Al ratio and acidity of mordenite

The Si/Al mol ratio in zeolites is ones of the most important character which affected the acidity, thermal stability, and activity in catalytic reactions. The dealumination process towards zeolite generally increases activity and thermal stability [17].

Based on the ICP measurement results, there was a change in the Si/Al ratio in the mordenite before and after the NaOH treatment (**Table 1**). The change in the Si/Al ratio indicates that the dealumination and desilication processes of mordenite are going well.

Based on **Table 1**, it shows an increase in Si/Al ratio of mordenite with the higher concentration of HNO_3. The increase in Si/Al ratio is due to the dealumination process in mordenite. In the dealumination process, previous experiment [18] stated

Mordenite code	Si/Al ratio	Total acid sites (mmol NH_3 g^{-1})
HM	9.8	3.67
$AM_{0.1}$	13.8	3.84
$AM_{0.5}$	14.1	2.27
BHM	8.8	4.24
$BAM_{0.1}$	14.8	4.19
$BAM_{0.5}$	15.3	2.85

Table 1.
The Si-Al ratio and acidity of mordenite after HNO3 and or NaOH treatment.

that the presence of H^+ ions resulting from the ionization of HNO_3 in water will cause H^+ ions to tend to be bound by O atoms that have bound Si and Al. This causes the O atom to have three unstable bonds. The stability of the O atom will be achieved if it breaks a bond that has the lowest dissociation energy. It can be seen that the dissociation energy of Al-O (116 kcal/mol) is much lower when it compared to the dissociation energy of Si-O (190 kcal/mol) so that the Al-O bond will be broken more easily than Si-O [19]. Therefore, the number of Al atoms in the framework will be reduced to non-framework Al due to the bond between Al atoms and NO_3^- ions.

The NaOH treatment was then carried out on mordenite HM, $AM_{0.1}$ and $AM_{0.5}$. This treatment can cause desilication which results in a decrease in the Si/Al ratio. In **Table 1**, it is shown that after NaOH treatment, Si/Al ratio of mordenite (AM) has decreased compared to that of mordenite (HM). The mordenite is in contact with NaOH which causes the silicon to be extracted from the mordenite framework, causing the Si/Al ratio decreased.

However, there were differences in mordenite $BAM_{0.1}$ and $BAM_{0.5}$, where there were an increase in Si/Al ratio compared to Si/Al ratio before being treated with NaOH as shown in **Table 1**. This is due to the low Si/Al ratio in mordenite before being treated with NaOH. The low Si/Al ratio causes of Si and Al distance to be close, so that when silicon is extracted, aluminum is also extracted due to its amphoteric nature [17].

The mordenite that has been treated with NaOH, then the ion exchange process is carried out with the ammonium ion. The exchange of cations in mordenite with NH_4^+ aims to remove impurities in the form of alkaline or cations such as K^+, Na^+, Ca^{2+}, Mg^{2+} which act as a counterweight to mordenite which can be exchanged with other cations so that the cations will be exchanged into NH_4^+ by pushing the cations. Then, the calcination process is carried out at high temperatures which aims to remove water molecules bound to the mordenite so that the surface area increases.

Besides the Si/Al ratio, another aspect that must be considered in mordenite is acidity. The calculation of the acid centre on the surface of a solid relates to the theory of Brønsted and Lewis acids, namely Brønsted acid (proton giver) and Lewis acid (an empty orbital capable of accepting an electron pair) present on the solid surface.

A simple method that can be used to determine acidity in solids or catalysts is by average of gravimetry, namely by weighing the solids before and after adsorbing the base. One of the bases that can be used as adsorbed substances is ammonia [20]. Ammonia base was chosen as the adsorbate because it has small molecular size so that it can be adsorbed into all mordenite pores.

Based on **Table 1**, it shows a decrease in the number of total acid sites (acidity) along with the increase in the Si/Al ratio of mordenite. The decrease in the total number of acid sites is due to a reduction in the number of Brønsted acid sites which present in the mordenite. The Brønsted acid site will decrease as the aluminum content in a zeolite decreases [4].

Increasing Si/Al ratio can lead the increase of acid strength. This is due to changes in the distribution of aluminum atoms in the zeolite framework. The farther of distance between the aluminum atoms, the less interference between the aluminum atoms, which causes an increase in the acid strength of the Brønsted acid sites possessed by aluminum atoms [21].

However, mordenite $AM_{0.1}$ was found to increase the number of acid sites when compared to mordenite HM. This could be due to the appearance of Lewis acid sites on the mordenite $AM_{0.1}$ framework. Lewis acid sites can occur if there are aluminum atoms bonded outside the main zeolite framework which is known as extra-framework aluminum (EFAL). This phenomenon also occurs in a study conducted by [6] who showed an increase in the number of acid sites on zeolites.

3.2 Effect of HNO_3 and/or NaOH treatment on the crystallinity of mordenite

Crystallinity of a sample solid can be analyzed using X-ray diffraction (XRD) which aims to determine and compare the crystallinity of mordenite either before or after treatment of HNO_3 and/or NaOH. The level of crystallinity of the catalyst can be seen from the peak intensity, while the type of mineral can be seen from the position of the diffraction angle (2θ) and the distance between planes. The crystallinity of a mordenite is very important, because it will be related to the stability of the mordenite. The results of mordenite analysis after HNO_3 treatment are shown in **Figure 1** and Intensity data of mordenite HM, $AM_{0.1}$ and $AM_{0.5}$ peaks in **Table 2**.

Table 2 shows a decrease in the crystallinity of mordenite in line with the increasing concentration of HNO_3, which can be seen in the decrease in mordenite intensity of $AM_{0.1}$ and $AM_{0.5}$ at 25.55°; 22.19°; and 27.51°. This was also confirmed by the percentage of crystallinity data of mordenite HM, $AM_{0.1}$ and $AM_{0.5}$ which were 100%, 91%, and 89%. This decrease in crystallinity was caused by the increasing amount of aluminum extracted from the mordenite (HM) framework [22].

The crystallinity test was also carried out on mordenite BHM, $BAM_{0.1}$ and $BAM_{0.5}$. The diffractograms of the three mordenites are shown in **Figure 2**. NaOH treatment has less effect on crystallinity than mordenite, where there was still a peak at 2θ 25.54°; 22.18°; and 27.50° and strengthened by mordenite intensity data BHM, $BAM_{0.1}$, and $BAM_{0.5}$ in **Table 3**.

Table 3 shows the intensity data possessed by the three mordenite catalysts at the 3 highest peaks. Based on these results, it can be seen that overall there is no significant change at the 2θ peak after NaOH treatment, this is also evidenced by the percentage data of mordenite crystallinity of 94%, 95%, and 94%.

Overall the crystallinity of mordenite treated with HNO_3 and/or NaOH did not change significantly. Therefore, this mordenite can be used further in the hydrolyzed α-cellulose hydrotreating process. This is because the hydrotreating process is carried out at high temperatures so that a catalyst with good crystallinity is needed or is maintained so that the stability of mordenite as a catalyst will also be maintained and will produce good activity.

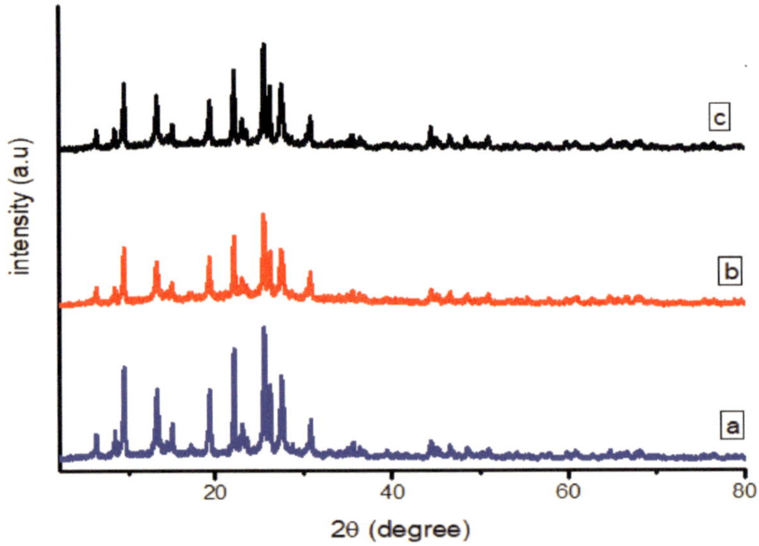

Figure 1.
Diffractogram HM (a), AM0.1 (b) and AM0.5 (c).

2θ (°)	Relative intensity		
	HM	$AM_{0.1}$	$AM_{0.5}$
25.55	1105	778	767
22.19	929	518	507
27.51	712	534	442

Table 2.
Intensity data of mordenite HM, AM0.1 and AM0.5 peaks.

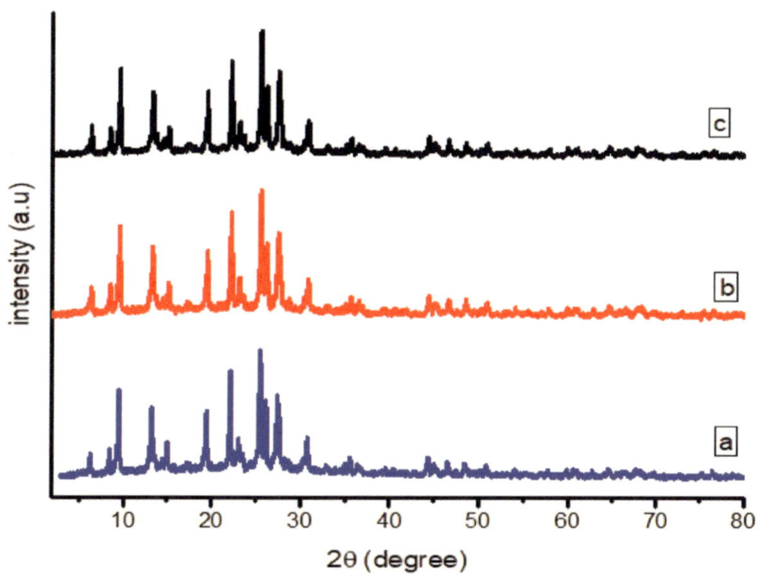

Figure 2.
Diffractograms of BHM (a), BAM0.1 (b) and BAM0.5 (c).

2θ (°)	Relative intensity		
	BHM	$BAM_{0.1}$	$BAM_{0.5}$
25.54	906	966	789
22.18	629	647	576
27.50	586	590	572

Table 3.
Intensity data of mordenite BHM, BAM0.1 and BAM0.5 peaks.

3.3 Effect of HNO_3 and/or NaOH treatment on mordenite pore character

The mordenite pore character analysis was performed using a Gas Sorption Analyzer (GSA). This pore character was carried out in order to determine the impact of the HNO_3 and/or NaOH treatment on the pore character of mordenite. The mordenite analyzed is mordenite that has been given alkaline treatment in order to determine the pore distribution of the mordenite. In addition, analysis by GSA provides information regarding the distribution of pore size (diameter), pore volume and specific surface area of mordenite. The specific surface area analysis of mordenite was carried out using the N_2 gas adsorption–desorption isotherm data on the mordenite surface based on the Branauer-Emmett–Teller (BET) method. The BET method describes the phenomenon of gas molecular adsorption on the surface of solids. N_2 gas is the most gas often used for analysis of the pore character of a material. The pore character in this study was carried out at a temperature of 77.35 K which is the boiling point of N_2.

Figure 3 below shows that the three mordenites have a type IV adsorption isotherm pattern. This type IV adsorption isotherm pattern indicates the formation of mesoporous characters in the three mordenites as indicated by the presence of loop hysteresis on the mordenite after being treated with NaOH. However, $BAM_{0.5}$ shows a wide and open hysteresis loop, it is presumed that in $BAM_{0.5}$ has pore diameter with the highest meso size compared to other mordenites and is followed by partial collapse of the mordenite framework, which is indicated by the presence of decrease in surface area drastically. Based on the shape of the hysteresis loop, it can be seen that the pore size formed in the three mordenites is homogeneous. This is supported by the pore distribution data obtained.

The pore distribution graph shows a relationship between dV (r) and pore diameter. Dv (r) (Differential volume radius) shows the pore distribution in mesoporous materials. The pore distribution in this study used the BJH method. Based on **Figure 4**, it can be seen that the pore distribution in the three materials

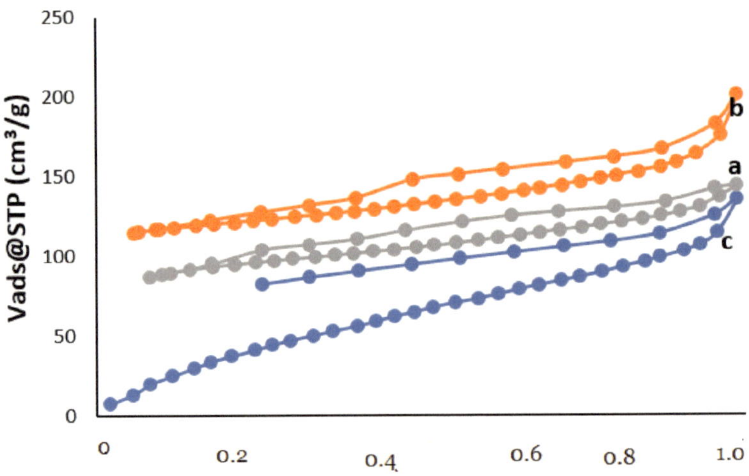

Figure 3.
Adsorption and desorption of isothermal gases N2 BHM (a), BAM0.1 (b) and BAM 0.5 (c).

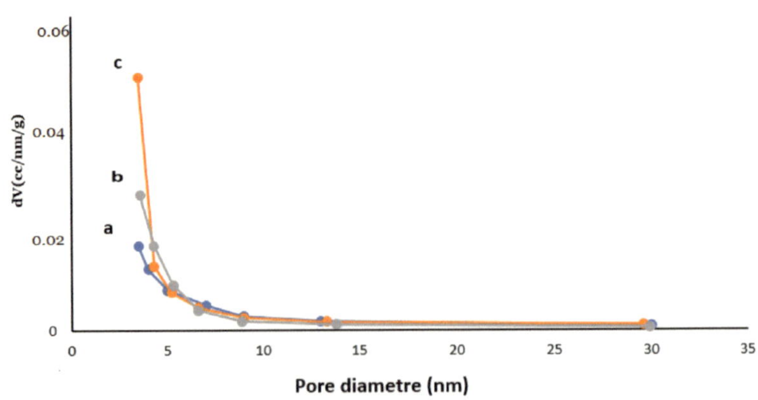

Figure 4.
Distribution of the desorption mordenite BJH BHM (a), BAM0.1 (b) and BAM0.5 (c).

is homogeneous. In addition, **Figure 4** also shows that all the pore distributions of the mordenite material are found in the mesoporous region, namely in the 3–30 nm range.

This is also supported by the average pore diameter data in **Table 4**. Based on the data in **Table 4**, it shows that alkaline treatment on mordenite can produce mordenite with a meso pore diameter. The higher the Si/Al ratio of mordenite, the resulting mesoporosity will also increase, this is because NaOH treatment can enlarge the pores than mordenite. The highest average pore diameter is owned by $BAM_{0.5}$, this is because the Si/Al ratio of BAM $_{0.5}$ is the highest compared to other mordenites.

The distribution of micropores and mesopores can be measured using a t-plot. This method can calculate the micropore volume and surface area of mesoporous rather than mordenite. The intercepts obtained from the t-plot method were 71.92 (BHM), 96.94 ($BAM_{0.1}$) and 13.03 ($BAM_{0.5}$). Through the intercept of the t-plot method, data on the pore volume composition of the mordenite will be obtained which are presented in **Table 5**.

Table 5 shows a decrease in the percentage of micro volume in mordenite after being treated with NaOH and followed by an increase in the percentage of volume of meso in the mordenite, along with the increase in the Si/Al ratio. This shows that the higher the Si/Al ratio of mordenite, the resulting mesoporosity will also increase.

The total surface area (S_{BET}) of mordenite was calculated using the multi-point BET method. Mesoporous surface area can also be calculated using the t-plot method. Mesoporous surface area calculations were performed using the slope of the mordenite t-plot chart. This calculation can produce data in the form of mesoporous surface area. The slopes obtained from the t-plot method are 5.259 (BHM), 5.445 ($BAM_{0.1}$) and 8.174 ($BAM_{0.5}$). Total surface area, micropore and mesoporous data are shown in **Table 6**.

Table 6 shows an increase in the percentage of the mesoporous surface area (S_{meso}) along with the increase in the Si/Al ratio of the mordenite. In addition, it was found a decrease in the total surface area of mordenite, but an increase in the percentage of Smeso. This is due to an increase in the surface area of the mesoporous structure compared to the micropore structure.

Mordenite code	Average pore diameter (nm)	Total pore volume (cm^3 g^{-1})	S_{BET} (m^2 g^{-1})
BHM	2.96	0.2223	371.5
$BAM_{0.1}$	3.34	0.3105	300.7
$BAM_{0.5}$	4.53	0.2101	185.6

Table 4.
Data of average pore diameter. Total pore volume and SBET.

Mordenite code	Total pore volume (cm^3 g^{-1})	V_{mikro} (cm^3 g^{-1})	V_{meso} (cm^3 g^{-1})	V_{mikro} (%)	V_{meso} (%)
BHM	0.2223	0.1113	0.1110	50.07	49.93
$BAM_{0.1}$	0.3105	0.1500	0.1605	48.31	51.69
$BAM_{0.5}$	0.2101	0.0202	0.1900	9.61	90.43

Table 5.
Data of Mordenite pore volume composition.

Mordenite code	S_{BET} (m^2 g^{-1})	S_{mikro} (m^2 g^{-1})	S_{meso} (m^2 g^{-1})	S_{mikro} (%)	S_{meso} (%)
BHM	371.5	290.1	81.36	78.09	21.90
$BAM_{0.1}$	300.7	216.5	84.23	72.00	28.01
$BAM_{0.5}$	185.6	59.64	126.0	32.13	67.89

Table 6.
Data on mordenite surface area composition.

3.4 Pyrolysis of α-cellulose (bio-oil)

Pyrolysis of α-cellulose in the form of a white powder produces a liquid product that has physical characteristics, dark brown, thick and has a strong odor. In addition, the oil resulting from α-cellulose pyrolysis can easily change color when left to stand at room temperature and open conditions. This change is possible due to the oxidation process. Experiment of [5] explained that bio-oil obtained from biomass pyrolysis is a multi-component mixture that has high acidity, high water content and is unstable in storage. The instability characteristic of the pyrolysis bio-oil is indicated by the easy color change of the α-cellulose pyrolysis bio-oil.

The liquid product resulting from α-cellulose pyrolysis was analyzed by GC–MS to determine the approximate compounds contained in the liquid product using data detected by MS after the liquid components were separated by the GC method. In addition, analysis of the content of the liquid product from α-cellulose pyrolysis by GC–MS is useful for knowing the main components of the liquid product from α-cellulose pyrolysis itself using area data from GC results. The results of the analysis of the liquid product content of α-cellulose pyrolysis are used as a reference in assessing the success of the hydrotreating process using a catalyst that has been synthesized. The conversion value of α-cellulose pyrolysis in this study was around 40–50% and the results of α-cellulose pyrolysis are shown in **Table 7**.

Table 7 shows that the liquid product resulting from α-cellulose is pyrolyzed containing furan, ketone, aldehyde, and carboxylic acid group compounds. Based on regional data, the five main components of the liquid product resulting from α-cellulose pyrolysis are 2-furancarboxaldehyde (21.21%), 1-hydroxy-2-propanone (20.96%), formic acid (9.15%), acetic acid (7.66%) and ethanal (7.27%). These results are the same as those of researchers [5] in their research report where the

Compound name	Molecular formula	Contains (%)
2-Furancarboxaldehyde	$C_5H_4O_2$	21.21
1-Hidroxy-2-Propanone	$C_3H_6O_2$	20.96
Formic acid	CH_2O_2	9.15
Acetic acid	$C_2H_4O_2$	7.66
Ethanal	C_2H_4O	7.27
5-Methyl-2-Furancarboxaldehyde	$C_6H_6O_2$	7.18
2-Propanone	C_3H_6O	5.73
1-Acetoxyethylene	$C_4H_6O_2$	3.22
2.3-Pentandione	$C_5H_8O_2$	2.92

Table 7.
Analysis results of the main product of a-cellulose by GC–MS.

liquid products from the hydrolyzed α-cellulose contained the main components in the form of furan compounds, ketones, carboxylic acids and aldehydes.

3.5 Catalyst activity

The catalyst activity test was carried out through the hydrotreating process. The hydrotreating process was carried out to determine the ability of the catalyst to convert the resulting α-cellulose from pyrolysis into a liquid fraction in the form of a more functional chemical. The analysis was carried out by determining the percentage of the liquid fraction produced as the main target, as well as the gas and coke fractions. The results of product distribution through the hydrotreating process can be seen in **Table 8**.

Based on **Table 8**, the catalytic hydrotreating process produces a higher liquid fraction than the liquid fraction without a catalyst (thermal). This is in accordance with the statement of [23] that the use of a catalyst in the hydrotreating process can increase the percentage of liquid fraction and product quality. Thermal hydrotreating and catalytic hydrotreating occur via different mechanisms. Thermal hydrotreating occurs through a free radical mechanism so that it is easier to produce products with short carbon chains and are gaseous in form. On the other hand, catalytic hydrotreating occurs through the carbonium ion mechanism which is easier to produce a liquid fraction with a long enough carbon chain.

The hydrotreating catalytic data in **Table 8** shows that the catalyst after being treated with a base produces a higher liquid fraction than the catalyst before being treated with a base. This is due to an increase in the pore distribution of higher meso size after alkaline treatment along with the increase in the Si/Al ratio of mordenite catalyst. However, the $BAM_{0.5}$ catalyst has a high mean pore diameter of meso compared to other catalysts, but produces a lower liquid fraction, this is due to a significant decrease in the number of acid sites at high Si/Al ratios. Therefore, the activity of a catalyst is not only determined by the pore diameter but also by the number of acid sites possessed by the catalyst. Increased activity in dealumination of mordenite was observed by [24], as well as [25], in a study using mordenite to convert m-xylene, where there was a decrease in m-xylene conversion in replicated mordenite due to a decrease in the concentration of the acid site.

3.6 Selectivity of liquid fraction

The liquid fraction resulting from the hydrolyzed α-cellulose hydrotreating was analyzed using a Gas Chromatography-Mass Spectrometer (GC–MS) to

Catalyst code	Product distribution% (b/b)		
	Liquid fraction	Gas fraction	Coke fraction
Thermal	23.16	76.48	—
HM	42.69	55.02	2.29
$AM_{0.1}$	59.73	36.62	3.65
$AM_{0.5}$	33.22	64.37	1.81
BHM	53.01	45.13	1.86
$BAM_{0.1}$	64.04	32.69	3.28
$BAM_{0.5}$	37.57	59.65	2.78

Table 8.
The results of product distribution through the hydrotreating process.

determine the selectivity of the catalyst. GC-MS analysis provides information from the hydrolyzed α-cellulose hydrotreating process. Selectivity is the tendency of a catalyst to produce certain components. The trends in the types of products produced in this study are functional chemicals.

The data in **Table 9** below can be used to explain the selectivity of catalyst use in the hydrolyzed α-cellulose hydrotreating process. The selectivity of the catalyst for the hydrotreating reaction can be seen based on the percentage of the main components of the resulting liquid fraction. The hydrotreating process both thermally and with a catalyst produces the main components of the same liquid fraction, namely acetic acid, 1-hydroxy-2-propanone, and ethanal, but with different liquid fraction percentages. But overall the hydrotreating products with mordenite after HNO_3 treatment still have the same compound as the pyrolyzed α-cellulose, these compounds are thought to be compounds from pyrolysis that do not convert to other compounds after hydrotreating.

Acetic acid, 1-hydroxy-2-propanone, and ethanal were produced with the highest product percentages of 11.03% (w/w), 11.49% (w/w), and 9.26% (w/w) on the use of an AM0.1 catalyst. The main components of the hydrolyzed α-cellulose hydrotreating liquid product in the form of acetic acid, 1-hydroxy-2-propanone (dihydroxyacetone) and ethanal are functional compounds that are widely used. Acetic acid is widely used as a food additive and is a very important chemical in various chemical industries. Acetic acid is an important industrial chemical used in the production of polyethylene terephthalate, cellulose acetate, polyvinyl acetate, and so on [26]. Study of [27] produced a 1-hydroxy-2-propanone compound which is used as a spice in food, colorants and additives in cosmetics. Ethanal is a compound that can be used as a solvent in the production of rubber, in tanning leather, in the paper industry, as a preservative for fruit and fish, and as an additive in flavorings in food.

The selectivity of the liquid fraction resulting from hydrotreating α-cellulose hydrolyzed using mordenite after NaOH treatment is shown in **Table 10**. **Table 10** shows that hydrotreating α-cellulose is hydrolyzed using mordenite after NaOH treatment to produce the main liquid fractions, namely ethanal and 1-hydroxy-2-propanone. Hydrotreating after NaOH treatment can produce hydrocarbon and alcohol group compounds that are not produced in hydrotreating by using mordenite after HNO_3 treatment. This is due to the formation of a mesoporous structure on the catalyst after NaOH treatment.

Group	**Product description**	**Bio-oil (%area)**	**Product contains % (b/b)**			
			Thermal	**HM**	**$AM_{0.1}$**	**$AM_{0.5}$**
Carboxylic acid	Acetic acid	7.66	3.50	7.19	11.03	4.59
	Formic acid	9.15	3.50	5.31		2.15
Ketone	1-Hidroxy-2-Propanone	20.96	4.67	10.43	11.49	6.89
Aldehyde	Ethanal	7.27	4.09	2.76	9.26	3.04
	2-Furancarboxaldehyde	21.21		2.63	5.25	2.45
Others		33.75	7.40	14.38	22.7	14.1
Total contains % (b/b)		100	23.16	42.69	42.69	33.22

Table 9.
Selectivity of hydrolyzed a-cellulose hydrotreating liquid fraction using mordenite after HNO3 treatment.

Group	Product description	Bio-oil (%area)	Product contains % (b/b)		
			BHM	$BAM_{0.1}$	$BAM_{0.5}$
Carboxylic acid	Acetic acid	7.66	8.33	9.56	
	Formic acid	9.15	2.44	5.33	
Ketone	1-Hidroxy- 2-Propanone	20.96	11.38	15.50	7.40
Aldehyde	Ethanal	7.27	6.72	7.05	4.26
	2-Furancarboxaldehyde	21.21	3.23		
Hydrocarbon	**1-Pentene**				**0.44**
	2-Heptyne				**2.75**
Alcohol	**1-Propanol**				**3.05**
Others		33.75	20.9	26.6	19.68
Total contain% (b/b)		100	53.01	64.04	37.57

Table 10.
Selectivity of hydrotreating a-cellulose liquid fraction by using mordenite after NaOH treatment.

Table 10 shows that the use of the $BAM_{0.5}$ catalyst in the hydrolyzed α-cellulose hydrotreating reaction gives the percentage of components of the hydrocarbon group, namely 1-pentene 0.44% (w/w) and 2-heptuna 2.75% (w/w) which is the fraction C5 and C7. The C5 and C7 fractions are included in the gasoline fraction where the gasoline fraction is a hydrocarbon that has a carbon chain range from C5-C12 [28], and produces an alcohol group compound, namely 1-propanol 3.05% (w/w). Joshi [29] stated that alcoholic fuels have been shown to increase the octane value and can be used as diesel and diesel fuel. Other catalysts as a whole only produce carboxylic acids, ketones, aldehydes and esters. This shows that $BAM_{0.5}$ has a Si/Al ratio, pore size, surface area, acidity, and has a good crystal structure and can work optimally in producing gasoline and alcohol fractions through the hydrolyzed α-cellulose hydrotreating process.

4. Conclusions

HNO_3 and/or NaOH treatments caused an increasing Si/Al ratio of mordenite HM with Si/Al ratio of 9.8 to 13.8; 14.1; 8.8; 14.8 and 15.3 which are $AM_{0.1}$, $AM_{0.5}$, BHM, $BAM_{0.1}$ and $BAM_{0.5}$, respectively. HNO_3 and NaOH treatments decreased the total acid sites from 3.67 to 3.84; 2.27; 4.24; 4.19 and 2.85 mmol NH_3 g^{-1} were $AM_{0.1}$, $AM_{0.5}$, BHM, $BAM_{0.1}$ and $BAM_{0.5}$, respectively and the treatments did not cause damage to the crystalline structure of the mordenite. NaOH treatment can produce mesoporosity in mordenite, which respectively have an average pore diameter of 2.96; 3.34 and 4.53 nm as BHM; $BAM_{0.1}$ and $BAM_{0.5}$. Catalyst $BAM_{0.5}$ has selectivity to 1-pentene 0.44% (w/w); 2-heptyne 2.75% (w/w) and 1-propanol 3.05% (w/w) from the hydrolyzed α-cellulose hydrotreating product.

Acknowledgements

The authors would like to thank The Ministry of Research, Technology and Higher Education, the Republic of Indonesia for the financial support under the scheme of PDUPT 2020.

Appendices and nomenclature

Appendices

$AM_{0.1}$	HM treated with HNO_3 0.1 M
$AM_{0.5}$	HM treated with HNO_3 0.5 M
BAM	acid treatment of mordenite
$BAM_{0.1}$	acid treatment of mordenite with HNO_3 0.1 M
$BAM_{0.5}$	acid treatment of mordenite with HNO_3 0.5 M
BHM	base/alkaline treatment of mordenite
HM	zeolite type mordenite (HSZ-604HOA)

Nomenclature

GC–MS	Gas Chromatography-Mass Spectrometer
GSA	Gas Sorption Analyzed
ICP	Inductively Coupled Plasma
XRD	X-Ray Diffraction

Author details

Wega Trisunaryanti*, Karna Wijaya, Desi Suryani and Uswatul Chasanah
Department of Chemistry, Gadjah Mada University, Yogyakarta, Indonesia

*Address all correspondence to: wegats@ugm.ac.id

References

[1] Ruh U, Mohammed ALHSS, Santiago S, Mert A. Adsorption Equilibrium Studies of CO_2, CH_4, and N_2 on Various Modified Zeolites at High Pressures Up to 200 Bars, Microporous and Mesoporous Materials. 2018;262:49-58. DOI: 10.1016/j.micromeso.2017.11.022

[2] Marli LG, Maura HJ Martin W Ernesto AUG. Synthesis of Mesoporous ZSM-5 by Crystallisation of Aged Gels in The Presence of Cetyltrimethylammonium Cations. Catalysis Today. 2008;133-135:69-79. DOI: https://doi.org/10.1016/j.cattod.2007.12.108

[3] Sònia A, Adriana B, Javier P. Mesoporous ZSM-5 Catalysts Prepared by Desilication with Organic Hydroxides and Comparison with NaOH Leaching, Applied Catalysis A: General. 2009;364: 191-198. DOI: https://doi.org/10.1016/j.apcata.2009.05.055

[4] Danny V, Nicholas N, Root L, Joost VA, Verolme P, Jeroen G, Javier P, and Sels B. Synthesis, Characterisation, And Catalytic Evaluation of Hierarchical Faujasite Zeolites: Milestones, Challenges, And Future Directions, Chemical Society Reviews, 2016;45(12):3331-3352. DOI: https://doi.org/10.1039/C5CS00520E

[5] Jun Zhao, Yanchao Y, Yang L, Wenyong C, Baijun L. Synthesis and Characterization of Mesoporous Zeolite-Y by Using Block Copolymers as Templates, Chemical Engineering Journal, 2016;284: 405-411. DOI: https://doi.org/10.1016/j.cej.2015.08.143

[6] Fuping TY, Yihui W, Shen Q, Xiang L, Yongying C, Changgong M. Effect of Si/Al Ratio on Mesopore Formation for Zeolite Beta Via NaOH Treatment and The Catalytic Performance in A-Pinene Isomerization and Benzoylation of Naphthalene, Microporous and Mesoporous Materials. 2013;173:129-138. DOI: https://doi.org/10.1016/j.micromeso.2013.02.021

[7] Johan CG, Tsuneji S, Jacob AM, Javier P. Alkaline-mediated Mesoporous Mordenite Zeolites for Acid-catalyzed Conversions, Journal of Catalysis.,2007;251: 21-27. DOI: https://doi.org/10.1016/j.jcat.2007.07.020

[8] Bo W, Chul WL, Tian-Xi C, and Sang-Eon P.Benzene Alkylation with 1-Dodecene Over H-Mordenite Zeolite, Catalysis Letters. 2001;76:1-2. DOI: https://doi.org/10.1023/A:1016776009518

[9] Pedro M, Jose MP, Pierre A, Laforge S, Magnoux P, Guisnet M, and Paulo FR. Effect of Deakumination by Acid Treatment of a HMCM-22 Zeolite on the Acidity and Activity of the Pore Systems, Applied Catalysis A: General , 2009; 365(2):207-213. DOI: https://doi.org/10.1016/j.apcata.2009.06.014

[10] Zahra VR, Shahla K, and Kharat AN. Dealumination of Mordenite Zeolite and Its Catalytic Performance Evaluation in m-Xylene Isomerization Reaction, Bulleti of the Chemical Society of Ethiopia. 2017;31(2):281-289. DOI: 10.4314/bcse.v31i2.9

[11] Nourrdine C, Ahcene S,Hussein I, Jean DC, and Ludovic P. Formation of Week and Strong Brønsted Acid Sites During Alkaline Treatment on MOR Zeolite, Applied Catalysis A: General. 2016;256:95-104. DOI: https://doi.org/10.1016/j.apcata.2016.08.004

[12] Wega T, Triyono, Karna W, Majid A.B, Priastomo Y, Febriyanti F, Hayyanti, dan Nugroho A. Karakterisasi dan Uji Aktivitas Katalis Mordenit dan Zeolit-Y Pada Hidrorengkah Ban Bekas menjadi Fraksi Bahan Bakar, Prosiding Seminar Nasional Kimia Unesa 2012-ISBN: 978-979-028-550-7.

[13] George WH, Sara I, and Avelino C. Synthesis of Transportation Fuels from Biomass: Chemistry, Catalysts, and Engineering. Chemical Reviews. 2006;106(9): 4044-4098. DOI: 10.1021/cr068360d

[14] Marika H and Tatjana B. Microwave-Supported Preparation of α-Cellulose for Analysis of δ13C in Tree Rings, Analytical Chemistry, 2006;78(20):7248-7252. DOI: 10.1021/ac060523a

[15] Tefvik A and Aimaro S. Nanno-chloropsis Algae Pyrolysis with Ceria-Based Catalysts for Production of High-Quality Bio-Oils, Bioresource Technology., 2015;194:108-116. DOI: https://doi.org/10.1016/j.biortech.2015.07.027

[16] Ramakanta S, Byung JS, Ji SI, Young-Pyo J, and Chul JW. A Review of Recent Advances in Catalytic Hydrocracking of Heavy Residues, Journal of Industrial and Engineering Chemistry, 2015;27:12-24. DOI: https://doi.org/10.1016/j.jiec.2015.01.011

[17] Marius CS, Céline C, and Pascal R. Challenger on Molecular Aspects of Dealumination and Desilication of Zeolites, Microporous Mesoporous Materials. 2014;191:82-96. DOI: http://dx.doi.org/10.1016/j.micromeso.2014.02.040

[18] Wega T, Sri S, dan Riyadi A. Pengaruh Perlakuan HCl terhadap Karakter Zeolit Alam Wonosari dan Uji Aktivitasnya untuk Perengkahan n-Heksadekana. Prosiding Seminar Nasional Kimia. 2005. UNS, Solo.

[19] Samuel G.Book of Physical Chemistry. 2nd Ed., John Weley and Sons; 1946. DOI: https://doi.org/10.1021/j150452a034

[20] Wega T, Ryuhei S, Masahiro, Masakatsu N,Norikazu N, and Masahiko M. Characterization and Modification of Indonesian Natural Zeolite and Their Properties for Hidrocracking of Parafin. Journal of The Japan Petroleum Institute. 1996; 1:20-25. DOI: 10.1627/jpi1958.39.20

[21] Jun H, Yijiao J, Reddy M, Bejoy T, Ekaterina R, and Michael H. Characterization and Acidic Properties of Aluminium-Exchanged Zeolites X and Y. Journal of Physical Chemistry. 2008;112(10): 3811-3818. DOI: https://doi.org/10.1021/jp7103616

[22] Edward FTL and Lovat VCR. Dealumination of Sodium Y Zeolite with Hydrochoric Acid, Journal of The Chemical Society. 1987;18:1531-1537. DOI: https://doi.org/10.1039/F19878301531

[23] Eelco V and Bert W. Fluid Ctalytic Cracking; Recent Developments on the Grand Old Lady of Zeolite Catalysis, Chemical Society Reviews. 2015; 44(20):7342-7370. DOI: https://doi.org/10.1039/C5CS00376H

[24] Aditya B and Enrique I. A Link Between Reactivity and Local Structure in Acid Catalysis on Zeolite, Accounts of The Chemical Research. 2008;41(4):559-567. DOI: https://doi.org/10.1021/ar700181t

[25] Paixão V, Monteiro R, Andrade M, Fernandes A, Rocha J, Carvalho AP, and Martins A. Desilication of MOR Zeolite: Conventional Versus Microwave Assisted Heating. Applied Catalyst A: General. 2011;402:59-68. DOI: https://doi.org/10.1016/j.apcata.2011.05.025

[26] Zheng S, Jingfei Z, Xuefei Z, and Yalei Z. The Production of Acetic Acid from Microalgae Under Hydrothermal Conditions. Applied Energy. 2011;88:3444-3447. DOI: https://doi.org/10.1016/j.apenergy.2010.12.060

[27] Mauricio V, Alexander S, and Catherine B. Conversion of Glycerol to Hydroxyacetone in Gas Phase Over La_2CuO_4 Catalyst. Applied Catalyst B: Environment 2014;160:606-613.

DOI: https://doi.org/10.1016/j.apcatb.2014.06.006

[28] Rodiansono and Wega T. Activity Test and Regeneration of NiMo/Z Catalyst for Hydrocracking of Waste Plastic Fraction to Gasoline Fraction. Indonesian Journal of Chemistry. 2005;5(3):261-268. DOI: https://doi.org/10.22146/ijc.21801

[29] Mayank CJ, Mayank J, Sumit S and Anurag BJ. Alcohol as an Alternative Fuel in I.C. Engine, International Journal of Engineering and Technical Research. 2016;4:2454-4698. DOI: https://www.erpublication.org/published_paper/IJETR041224.pdf